# SOFTSWITCH

# MCGRAW-HILL NETWORKING

**Demystified**

| | |
|---|---|
| Harte/Levine/Kikta | *3G Wireless Demystified* |
| LaRocca | *802.11 Demystified* |
| Muller | *Bluetooth Demystified* |
| Hershey | *Cryptography Demystified* |
| Hoffman | *GPRS Demystified* |
| Symes | *MPEG-4 Demystified* |
| Camarillo | *SIP Demystified* |
| Topic | *Streaming Media Demystified* |
| Symes | *Video Compression Demystified* |
| Shepard | *Videoconferencing Demystified* |

**Developer Guides**

| | |
|---|---|
| Guthery | *Mobile Application Development with SMS* |
| Richard | *Service and Device Discovery: Protocols and Programming* |

**Professional Telecom**

| | |
|---|---|
| Smith/Collins | *3G Wireless Networks* |
| R. Bates | *Broadband Telecom Handbook,* Second Edition |
| Collins | *Carrier Class Voice over IP* |
| Minoli | *Ethernet-Based Metro Area Networks* |
| R. Bates | *GPRS* |
| Minoli | *Hotspot Networks Wi-Fi for Public Access* |
| R. Bates | *Optical Switching and Networking Handbook* |
| J. Bates | *Optimizing Voice in ATM / IP Mobile Networks* |
| Sulkin | *PBX Systems for IP Telephony* |
| Russell | *Signaling System #7, Fourth Edition* |
| Saperia | *SNMP at the Edge: Building Service Management Systems* |
| Minoli | *Voice over MPLS* |
| Karim/Sarraf | *W-CDMA and cdma2000 for 3G Mobile Networks* |
| R. Bates | *Wireless Broadband Handbook* |
| Faigen | *Wireless Data for the Enterprise* |
| Nichols | *Wireless Security* |

**Reference**

| | |
|---|---|
| Muller | *Desktop Encyclopedia of Telecommunications, Third Edition* |
| Botto | *Encyclopedia of Wireless Telecommunications* |
| Clayton | *McGraw-Hill Illustrated Telecom Dictionary, Third Edition* |
| Radcom | *Telecom Protocol Finder* |
| Kobb | *Wireless Spectrum Finder* |

# Softswitch
## Architecture for VoIP

Franklin D. Ohrtman, Jr.

**McGraw-Hill**

New York   Chicago   San Francisco   Lisbon
London   Madrid   Mexico City   Milan   New Delhi
San Juan   Seoul   Singapore   Sydney   Toronto

**The McGraw-Hill Companies**

**Cataloging-in-Publication Data is on file with the Library of Congress.**

Copyright © 2003 by The McGraw-Hill Companies, Inc. All rights reserved.
Printed in the United States of America. Except as permitted under the United
States Copyright Act of 1976, no part of this publication may be reproduced or
distributed in any form or by any means, or stored in a data base or retrieval
system, without the prior written permission of the publisher.

3 4 5 6 7 8 9 0  DOC/DOC  0 9 8 7 6 5 4

ISBN 0-07-140977-7

*The sponsoring editor for this book was Marjorie Spencer and the production supervisor
was Pamela A. Pelton. It was set in New Century Schoolbook by MacAllister Publishing
Services, LLC.*

*Printed and bound by RR Donnelley.*

McGraw-Hill books are available at special quantity discounts to use as premiums and
sales promotions, or for use in corporate training programs. For more information,
please write to the Director of Special Sales, Professional Publishing, McGraw-Hill,
Two Penn Plaza, New York, NY 10121-2298. Or contact your local bookstore.

 This book is printed on recycled, acid-free paper containing a minimum of
50 percent recycled de-inked fiber.

# CONTENTS

Foreword                                                                    xiii
Preface                                                                      xv
Acknowledgments                                                             xvii

**Chapter 1**   Introduction                                                 1

Softswitch as an Alternative to Class 4 and Class 5                          4
   Reliability                                                               4
   Scalability                                                              5
   Quality of Service (QoS)                                                 5
   Signaling                                                                5
   Features                                                                 6
   Regulatory Implications                                                  6
Economic Advantage of Softswitch                                           7
Disruptive or Deconstructive Technology?                                   7

**Chapter 2**   The Public Switched Telephone Network (PSTN)                9

Access                                                                     10
Switching                                                                  11
   Class 4 and 5 Switching                                                 12
   Private Branch Exchange (PBX)                                           14
   Centrex                                                                 15
   Multiplexing                                                            16
   Voice Digitization via Pulse Code Modulation                           16
   Signaling                                                               21
   The Advanced Intelligent Network (AIN)                                  27
   Features                                                                29
   Performance Metrics for Class 4 and 5 Switches                         30
Transport                                                                  34
   Asynchronous Transfer Mode (ATM)                                        34
   Optical Transmission Systems                                            35
Conclusion                                                                 38

**Chapter 3**   Softswitch Architecture or "It's the Architecture, Stupid!"   39

Softswitch and Distributed Architecture: A "Stupid" Network                40
Access                                                                     42
   PC to PC and PC to Phone                                                42
   IP Phones (IP Handsets) Phone-to-Phone VoIP                            43
   Media Gateways (a VoIP gateway switch)                                  46

Switching                                                            50
    Softswitch (Gatekeeper and Media Gateway Controller)          50
    Signaling Gateway                                             52
    Application Server                                            52
    Applications for Softswitch                                   53
    Class 4 Replacement Softswitch                                56
    Class 5 Replacement Softswitch                                58
Transport                                                          60
    Legacy, Converging, and Converged Architecture                60
    IP Networks                                                   61
    ATM                                                           63
    TDM                                                           64
Conclusion                                                         64

**Chapter 4**   Voice over Internet Protocol                       67

What Is VoIP?                                                      68
Origins                                                            68
How Does VoIP Work?                                                69
Protocols Related to VoIP                                          70
    Signaling Protocols                                           71
    Routing Protocols                                             79
    Transport Protocols                                           84
IPv6                                                               85
Conclusion                                                         86

**Chapter 5**   SIP: Alternative Softswitch Architecture?          87

What Is SIP?                                                       89
    SIP Architecture                                              89
    New Standards for SIP                                         96
Some SIP Configurations                                            97
Comparison of SIP to H.323                                         98
Complexity of H.323 Versus SIP                                     99
    Scalability                                                  102
    Extensibility                                                105
    Services                                                     108
    H.323 Versus SIP Conclusion                                  109
The Big "So What?" about SIP                                      109
    SIP on Windows XP                                            109
    How Does That Work?                                          110
Conclusion                                                       112

| | | |
|---|---|---|
| **Chapter 6** | Softswitch: More Scalable Than CLASS 4 or 5 | 113 |
| | Scalability | 114 |
| | Scaling Up | 115 |
| | Scaling Down | 119 |
| |    Scaling Down for Class 4 Applications | 120 |
| |    Scaling Down for Class 5 or Central Office Bypass Applications | 121 |
| |    Scaling Down for Class 5 or Central Office Bypass from the Residence | 122 |
| |    Access Switching | 124 |
| |    Scaling Down for Class 5 or Central Office Bypass from the Enterprise | 126 |
| |    Scaling Down Technical Issues | 127 |
| | Conclusion | 128 |
| | | |
| **Chapter 7** | Softswitch Is Just as Reliable as Class 4/5 Switches | 131 |
| | World Trade Center Attack: A Need to Redefine Reliability | 132 |
| |    One to Five 9s | 135 |
| |    Standards for Availability | 135 |
| | What Is Reliability? | 135 |
| |    How Availability Is Calculated | 137 |
| | How Does a Switch, PSTN or Softswitch, Achieve Five 9s? | 138 |
| |    Network Equipment Building Standards (NEBS) | 144 |
| | Specifications for Softswitch Reliability | 145 |
| | Software Reliability Case Study Cisco IOS | 147 |
| | Power Availability | 149 |
| |    Typical Power Outages in a Typical Telephone Network | 149 |
| | Human Error | 150 |
| | Conclusion | 152 |
| | | |
| **Chapter 8** | Quality of Service (QoS) | 155 |
| | Factors Affecting QoS | 156 |
| | Improving QoS in IP Routers and the Gateway | 157 |
| |    Sources of Delay: IP Routers | 157 |
| |    Sources of Delay: VoIP Gateways | 158 |
| |    Other Gateway Improvements | 159 |
| |    Perceptual Speech Quality Measurement (PSQM) | 160 |
| | Improving QoS on the Network | 161 |
| |    Resource Reservation Setup Protocol (RSVP) | 162 |
| |    Differentiated Service (DiffServ) | 165 |

MPLS-Enabled IP Networks                                    167
MPLS Architecture                                           170
MPLS Traffic Engineering                                    171
Measuring Voice Quality                                     173
Mean Opinion Score (MOS)                                    173
Conclusion                                                  174

**Chapter 9**   SS7 and Softswitch                          175

Signaling in the PSTN (SS7 or C7)                           178
Message Transfer Part (MTP)                                 178
ISDN User Part (ISUP)                                       179
Signaling Connection Control Part (SCCP)                    179
Transaction Capabilities Applications Part (TCAP)           179
Interworking SS7 and VoIP Networks                          180
Signaling in VoIP Networks                                  181
Signaling Transport (SIGTRAN)                               182
SIGTRAN Protocols                                           183
Stream Control Transmission Protocol (SCTP)                 184
Transporting MTP2 over IP: M2UA                             186
ISDN Q.921-User Adaptation Layer (IUA)                      189
Signaling Network Architecture                              190
SS7 Interworking with SIP and H.323                         191
ISUP Encapsulation in SIP                                   191
SIP for Telephones (SIP-T)                                  192
PINT and SPIRITS                                            192
Interworking H.323 and SS7                                  193
Conclusion                                                  194

**Chapter 10**  Features and Applications: "It's the Infrastructure, Stupid!"   195

Features in the PSTN                                        196
Features and Signaling                                      198
The Intelligent Network/Advanced Intelligent
Network (IN/AIN)                                            198
Service Creation Environment                                200
Application Programming Interfaces (APIs)                   202
APIs and Services                                           202
SIP as API: International Softswitch Consortium's Architecture
for Enhanced Services in a Softswitch Network               208
Media Servers                                               209

# Contents

Application Servers 209
Architecture 209
Physical Architecture 211
Interface Between Call Control and Application Servers 212
Application Server to Media Server Interface 216
Service APIs 216
Application Server Interactions 216
Application and Media Servers Summary 217
How Softswitch Handles E911 and CALEA Requirements 218
Enhanced 911 (E911) 218
Communications Assistance for Law Enforcement Act (CALEA) 219
Next-Generation Applications Made Possible by Softswitch
Features 221
Web Provisioning 221
Voice-Activated Web Interface 222
"The Big So What?" of Enhanced Features 222
Napster and i-Mode: Examples of Killer Apps 222
Conclusion 223

**Chapter 11**  Softswitch Economics 225

A Previous Example of Disruptive Technology in the
Long-Distance Industry 226
Softswitch Is More Cost Effective Than Class 4 229
Purchase and OAM&P 230
Bandwidth Saving 231
Lower Barrier to Entry 232
Softswitch: A Smaller Footprint 233
Softswitch Advantages in Power Draw 236
Advantages of Distributed Architecture 237
Economic and Regulatory Issues Concerning Softswitch 238
Net Present Value of Softswitch 239
Considerations for the Net Present Value Models 241
Net Present Value: Midsized Long-Distance Service Provider 249
Leasing Class 4 Versus Softswitch 250
Buying Class 4 or Softswitch 4,032 DS0s 252
Buying Class 4 or Softswitch with Class 4 at $50 per DS0 253
Net Present Value When Softswitch Generates Greater Revenue 255
Large Long-Distance Service Providers 257
Summary of Net Present Value Analyses 259

$0 per port: Subscriber Pays for Access Device Negating
   "Cost per Port"                                                                      261
Economics of Enterprise Softswitch Applications                                                       262
Non-Interoffice Free Long Distance                                                                    262
Implications for Developing Economies                                                                 265
Economic Benefits of Converged Networks                                                               266
   Broadband Access and Telephone Services                                              267
Conclusion                                                                                            269

**Chapter 12**  Is Softswitch Deconstructive, Disruptive, or Both?                           271

Deconstruction                                                                                        272
   Deconstruction of Service Providers                                                  273
   Vendors                                                                              276
Disruption of the Legacy Telecommunications Value Network                                             277
   Disruption in Class 4 Market                                                         277
   Disruption in PBX Market                                                             284
Conclusion                                                                                            285

**Chapter 13**  Softswitch and Broadband                                                     287

Converged Networks Independent of ILEC Infrastructure:
   "We Have the Technology"                                                              288
Access Alternatives to the PSTN and Cable TV                                                          289
   Wi-Fi (802.11b Standard)                                                             289
Fiberless Optics                                                                                      290
$500 Billion Economic Benefit of Converged Broadband
   Networks                                                                             291
Broadband Access and Telephone Services                                                               292
National Defense Residential Broadband Network (NDRBN)                                                 295
Better Living Through Telecommunications: The Social
   Rewards of Softswitch                                                                296
   Essay One - If It Hurts to Commute, Then Don't Commute                                296
   Essay Two: Affordable Housing Is Where You Find It                                    299
   Essay Three: Family Values                                                           300
The Role of Softswitch in Better Living Through
   Telecommunications                                                                   300
Conclusion                                                                                            302

# Contents

**Chapter 14** Past, Present, and Future of Softswitch 303

History of Softswitch 304
The Present of Softswitch: Case Studies 305
    Class 4 Replacing the Class 4 Switch in Long-Distance
        Applications: Fusion, Sonus Networks, and NexVerse 305
    Replacing the Class 5 Switch: NorVergence and MetaSwitch 306
The Future of Softswitch: ISC Reference Architecture and the ISC 307
Functional Planes 309
    Transport Plane 310
    Call Control & Signaling Plane 311
    Service & Application Plane 311
    Management Plane 311
Functional Entities 311
    Media Gateway Controller Function (MGC-F) a.k.a. Call
        Agent or Call Controller 312
    Call Routing and Accounting Functions (R-F/A-F) 313
    Signaling Gateway Function (SG-F) and Access Gateway
        Signaling Function (AGS-F) 314
    Application Server Function (AS-F) 315
    Media Gateway Function (MG-F) 316
    Media Server Function (MS-F) 317
Media Gateway Controller Building Blocks 318
    MGC Implementation Example 319
Network Examples 320
    Wireline Network 320
    All-IP Network 322
    VoIP Tandem Switching 323
    POTS Carried over IP 324
    Access Network (V5/ISDN) over IP 324
Cable Network (e.g. PacketCable™) over IP 326
    VoDSL and IAD over IP 327
    Wireless (3GPP R99 Special Case NGN) 327
    Wireless (3GPP R2000 General Case all IP) 328
    WCDMA Mobile Network 328
Conclusion 329

**Chapter 15** Conclusion and Prognostications 331

Softswitch and the PSTN Metrics of Performance 333
Softswitch and the PSTN Infrastructure 334
Alternatives to the Telephone Company 335
Alternative Private Service Providers 336
Alternative Public Service Providers 336
The End of the PSTN as We Know It? 337

Acronyms List 339

Index 345

# FOREWORD

What is a softswitch? Softswitch technology is an enabling platform for next-generation packet communications including voice, broadband, and wireless networks. Softswitch technology and applications enable global service providers and carriers to optimize their networks and generate new revenue streams with new services and applications.

The telecommunications industry is now going through a cycle that economist Josef Schumpeter would call "creative destruction." The legacy infrastructures and business models, like organic entities, have gone into decline and must be replaced. The technologies, business practices, and services of the incumbent telephone companies are outdated and cannot meet the consumer demands of the future. Softswitch technology falls into what Professor Christensen of the Harvard Business School calls "disruptive" because it is demonstrably "cheaper, simpler, smaller, and more convenient to use." Softswitch could potentially disrupt the status quo of the telecommunications industry as we know it.

In order for softswitch to be a viable platform in the telecommunications industry, it must meet or exceed the performance parameters of legacy Time Division Multiplexing (TDM) switches before service providers, incumbents, or new market entrants will trust that technology in their networks. Those performance parameters include scalability, reliability, quality of service (QoS), features, and signaling capabilities.

This book illustrates how softswitch technology disrupts incumbent service providers and their vendors. It helps to identify some of the success and technological challenges of the industry based on the softswitch.

Michael H. Khalilian, Telecom Executive and Analyst
m.khalilian@attbi.com

# PREFACE

At the time of this writing, the sky is falling in the telecommunications industry. Bankruptcies of telecommunications companies have reached historical proportions. What we are witnessing is a process of what Joseph Schumpeter called Creative Destruction, an essential element of capitalism. Free market institutions (telephone companies and their suppliers in this case) are being destroyed as Darwinian competition continually forces them to change to match or exceed the competitors' performance.

With this process in mind, this book focuses on the metrics of performance of the emerging telecommunications infrastructure in terms of scalability, reliability, quality of service (QoS), signaling, and features. By analyzing how softswitch technologies might meet or exceed the performance parameters of the PSTN infrastructure (specifically Class 4 and 5 switches in addition to PBXs), it becomes increasingly clear how Creative Destruction takes place from an engineering perspective.

The criteria for comparison were selected by the contemporary telecommunications industry itself. These are performance criteria service providers' demand of vendors. If a platform exceeds the performance criteria of the legacy infrastructure it may be adopted by a service provider in order to head off competition in its selected market. Failure to adapt to changing market conditions via "cheaper, simpler, smaller and more convenient," to quote Clayton Christiensen, technologies leads to disruption, that is, the demise of what are otherwise well-run companies.

Flexibility in scalability is one aspect of softswitch technologies where the inability to adopt softswitch technologies can severely impair a service provider's ability to compete in the telecommunications market. For the last few decades, the industry has followed a credo of "bigger is better" in regard to switch capacity (number of DS0s and BCHAs), which is in direct contradiction to Christensen's argument that smaller packages are disruptive to the status quo.

Service providers and their suppliers have long believed that their switch technologies were unassailably dependable (reliable). Yet switch technologies deployed around the world are based on 1970s mainframe computer technology. The "five 9s" of a switch designed in the 1970s is woefully inadequate for the demands of the 21st century where the whole network, and not just its switches, must deliver "five 9s" of reliability. The events of September 11, 2001 are powerful testimony of the unreliability of legacy telephone networks.

Skeptics of softswitch technologies once made the charge that softswitch was not a viable technology, as it did not deliver the 3,500 features of a Class 5 switch. The delivery of an almost unimaginable array of new services is made possible by the introduction of IP technologies into voice communications. Not only does the use of IP greatly simplify signaling, it introduces a whole new array of services not possible with legacy switches.

Some part of the explanation for the financial predicament for telecom vendors is that softswitch technology is cheaper in both acquisition and operation. Service providers have traditionally depreciated telephone switches over a period of tens of years. Given the promise of softswitch, many service providers, both incumbent and new market entrant, are either buying softswitch platforms or are waiting for softswitch as a technology to gain more experience in the marketplace. Sales of legacy switches are depressed as a result. The lower costs of acquisition and operation of softswitch poses a lower barrier to entry for new market companies. The rise in the number of market players dilutes the amount of market share enjoyed by incumbents and drives competition in the marketplace. Softswitch enables any provider of IP services to transmit voice.

Softswitch is a rapidly maturing industry. The Internet Engineering Task Force (IETF), the International Telecommunications Union (ITU), and the International Softswitch Consortium (ISC) have individually composed a raft of standards and guidelines for VoIP and softswitch technologies. Service providers need not fear a gross lack of standardization in this industry.

# ACKNOWLEDGMENTS

This book was made possible with the generous assistance of the following persons: Scott Savage, John Thompson, Robert Mercer, Dale Hatfield, Jonathan Rosenberg, Henning Schulzrenne, Thom Baker, Bertha Latorre, Juan Jose Furukawa, Julian Martinez, Michael Khalilian, Nathan Stratton, Jonathan Christianson, Monica Kicklighter, Joan Lockhart, David Isenberg, Dale Hartzell, Ike Elliott, Lily Sun, Rachel Shelton, Scott Hoffpauir, Rod Beck, Tony Downes, Dharma Kuthanur, Debbie O'Brien, Mike Hluchyj, John Breeding, William Herwig, Mauricio Arriaga, Carlos Elizondo, Edgar Brizuela, Michael Writer, Clayton Christensen, Gurshuran Sidhu, Scott Anthony, Erik Roth, Paul Waibel, David Butler, Mat Osman, Andrew Snow, Charles Jackson, Robert Crandall, Marjorie Spencer and Iain Stewart.

A special acknowledgment goes to Eric Burger, Chief Technology Officer of SnowShore Networks. Mr. Burger's vast experience and contribution as technical editor for this book proved absolutely invaluable.

A special acknowledgment goes out to Heather Thompson, Robin Balchen, and Jennifer LeGrand, the morning barristas at St. Mark's Coffee House in Denver, Colorado without whose coffee and kind smiles, this book would not have been possible.

SOFTSWITCH

CHAPTER 1

# Introduction

In 2000, the telecommunications boom went bust, and the reason was that new market entrants, known as *Competitive Local Exchange Carriers* (CLECs), were forced to compete with *Incumbents Local Exchange Carriers* (ILECs) on the terms of the incumbents. The failure of the CLECs resulted in a net investment loss of trillions of dollars, adversely affecting capital markets and severely depressing the overall telecommunications economy, as well as saddling subscribers with artificially high rates. The chief expense for a new market entrant was purchasing and maintaining one or more Class 5 switches (for local service providers) or Class 4 switches (for-long-distance service providers). These switches cost millions of dollars and came with expensive maintenance contracts. They were also very large and required expensive central office space. Faced with competing for thin margins on local telephone service or thinner long-distance margins against incumbents who enjoyed strong investor support and long depreciation schedules on capital equipment, the demise of many new market entrants was foretold by their balance sheets.

The Telecommunications Act of 1996 aimed to introduce competition into the local loop by legally requiring the incumbents to lease space on their switches and in their central offices to any and all competitors. These regulations failed to have the desired effect, for reasons that have been meticulously analyzed and are still being debated today. The mixture of unincentivized telcos, courts without case law, and startups without capital and influence proved deadly.

Given firstly the astronomical expense of buying and installing Class 4 or 5 switches and secondly the legal obstacle of gaining access to the *Public Switched Telephone Network* (PSTN), it is little wonder that six years after the passage of the Telecommunications Act of 1996 only nine percent of American residential phone lines are handled by competitive carriers. Six years after the passage of the Act, 91 percent of all American households have, as their choice of telephone service provider, the Regional Bell Operating Company or the Regional Bell Operating Company.

A competitive local loop environment has two apparently insurmountable obstacles: (1) the high cost of Class 4 and 5 switches and (2) access to the local loop network. By 2002, some service providers had come to believe that competition will never come *in* the local loop but will have to come *to* the local loop in the form of an alternative network. Clearly a competitive network based on Class 4 and 5 switches would be cost-prohibitive. The only way consumers will enjoy the benefits of competition in the local loop, providers contend, is via alternative technology in switching and eventually in access, to lower the bar on entry and exit.

The primary access problem for CLECs is in the network between the central office (where Class 5 and 4 switches are located) and the residence or business. Although a variety of wires provides access to a residence (telephone, cable TV, and electrical) and wireless telephone service has exploded in popularity worldwide, voice services until recently had no technology capable of bypassing the central office. There was, in short, no alternative switching architecture to capitalize on alternative means of access (cable, wireless, and so on).

The lack of competition in and to the local loop raises the specter of another problem inherent in a monolithic telecommunications structure. What happens when major hubs of the PSTN are destroyed in natural disasters, terrorist attacks, or other force majeurs? The September 11th attack on the World Trade Center has served to focus attention on the vulnerability of the legacy, circuit-switched telephone network. Verizon, the largest telephone company affected, had five central offices that served some 500,000 telephone lines south of 14th Street in Lower Manhattan. More than six million private circuits and data lines passed through switching centers in or near the World Trade Center. AT&T and Sprint switching centers in the WTC were destroyed. Verizon lost two WTC-specific switches in the towers, and two nearby central offices were knocked out by debris, fire, and water damage. Cingular Wireless lost six towers and Sprint PCS lost four. Power failures interrupted service at many other wireless facilities.[1] Verizon further estimates 300,000 voice business lines, 3.6 million data circuits, and 10 cellular towers were destroyed or disrupted. This equates to phone and communications service interruption for 20,000 residential customers and 14,000 businesses.[2]

Business and residential customers of these service providers had no backup. Some were without service for weeks after the disaster. Financial losses from this extended outage are incalculable. Another casualty was the public's untested belief that the PSTN is invincible.

The American PSTN can be described as having a centralized architecture. The telephone companies have not built redundancy into their networks. Almost all cities and towns across the nation rely on one hub or central office, meaning that if a hub were destroyed, that city would lose all land-line telephone connectivity with the outside world. Even among CLECs, fewer than 10 percent have facilities truly separate from the RBOCS. Moreover, between 1990 and 1999, the number of RBOC central

---

[1]Telecom Update #300, September 17, 2001, www.angustel.ca/update/up300.html.
[2]Naraine, Ryan. "Verizon Says WTC Attacks May Hurt Bottom Line," *Silicon Alley News*, www.atnewyork.com/news/article/0,1471,8471_897461,00.html.

offices grew less than one percent to a nationwide total of 9,968, while the total number of phone lines grew by 34 percent according to the FCC.[3]

This trend toward a more centralized infrastructure has significantly elevated the cost of any catastrophic outage at the central office, without in any way mitigating the risk. It is widely believed that the introduction of an alternative network infrastructure could reduce both the risk of, and the cost incurred by, a PSTN failure.

# Softswitch as an Alternative to Class 4 and Class 5

Although too late for the failed new market entrants of the telecom boom of the late 1990's, new technologies have arrived on the market that provide a low-cost alternative to Class 4 and 5 switches in both purchase price and cost of maintenance. These technologies are *Voice over Internet Protocol* (VoIP) and softswitch. Softswitch provides the call control or intelligence for managing a call over an *Internet Protocol* (IP) or other network. Would these technologies have stayed off the bust? No one knows if they can replicate those qualities of Class 4 and 5 switches that made them industry standards for the last 25 years. Those qualities are reliability, scalability, *quality of service* (QoS), features, and signaling. Many have argued that VoIP and softswitch technologies must match Class 4 and 5 switches in all of these qualities before their deployment in a market environment is feasible.

That time may have come. Not only do VoIP and softswitch compare favorably in function and quality with Class 4 and 5 switching, but they deliver services not possible with Class 4 and 5 switches. Novel services could potentially generate additional revenues for providers, making them at least as profitable as the incumbents and introducing true competition to the local loop.

## Reliability

Reliability is plainly the chief point of comparison for service providers evaluating competitive technology. Class 4 and 5 switches have a reputation for

---

[3]Young, Shawn, and Dennis Berman. "Trade Center Attack Shows Vulnerability of Telecom Network," *Wall Street Journal*, October 19, 2001, p.1.

the "five 9s" of reliability. That means they will be in service 99.999 percent of the time. Building a voice-switching solution to achieve five nines is neither black magic nor a mandate from heaven on golden tablets. It is a matter of meticulously engineering into the solution the elements of redundancy, no single point of failure, and *Network Equipment Building Standards* (NEBS). Many softswitch solutions already offer "five 9s" or better reliability.

## Scalability

Scalability is also key to service providers. To compete with a Class 4 or 5 switch, a softswitch solution must scale up to 100,000 DS0s (for example, phone lines or ports). On this score, softswitch solutions excel. By virtue of new, high-density media gateways, they now match 100,000 DS0s in a single 7-foot rack, as opposed to the 39 racks typical of a Class 4 or 5 switches. And scalability is bidirectional. Softswitches can scale down to as little as two-port media gateways or even one port in the case of IP handsets, allowing unlimited flexibility in deployment. The minimum configuration for a Class 4 switch, for example, is 480 DS0s.

## Quality of Service (QoS)

Early VoIP applications garnered a reputation for poor QoS. First available in 1995, these applications were often little more than personal computers with microphones and speakers transmitting packets over the public Internet. Calls were often dropped and voice quality was questionable. Vast improvements in IP networks over the last seven years coupled with advances in media gateway technologies now deliver the potential for a QoS to match that delivered via Class 4 and 5 switches over the PSTN.

## Signaling

One element of the PSTN that was designed to deliver good QoS and thousands of features is *Signaling Service 7* (SS7). The interfacing of SS7 and IP networks necessary to deliver calls that travel over both the PSTN and an IP network is a significant challenge. Much progress has been made, recently including the emergence of a new technology designed to operate with IP networks known as SigTran. SigTran emulates SS7. In addition, the VOIP industry has new protocols such as the *Session Initiation Protocol* (SIP) that, although very different from SS7, match its signaling capabilities.

## Features

Proponents of the PSTN can challenge VoIP and softswitch solutions with a peremptory "Where's the 3,500 5ESS features?" This question refers to Lucent Technologies *#5 Electronic Switching System* (5ESS) Class 5 switch, long reported to have that many calling features. Proponents of softswitching question in turn whether each and every one of those 3,500 features is necessary to the successful operation of a competitive voice service. Today, telcos that require new features must contract with the switch vendor to obtain them. Providing those features will take months of development and hundreds of thousands of dollars.

By contrast, softswitch solutions are often based on open standards and use software applications such as *Voice XML* (VXML) to write new features. Service providers using softswitch solutions can often write their own features in house in a matter of days. They can also obtain new features from third-party software vendors. Given this ease and economy of developing custom features, the question of how many features are already on the menu becomes less compelling.

This flexibility in deploying new features offers service providers a first-to-market edge for high-margin features over Class 4 or 5 switch solutions. In a Net Present Value calculation, a softswitch solution, given its lower cost of acquisition and operation coupled with an ability to generate greater revenues, will prevail over a Class 4 or 5 solution.

## Regulatory Implications

The regulatory environment in the American telecommunications market is sympathetic to VoIP and softswitch solutions. Long-distance VOIP calls in the United States are immune to access fees and *Universal Service Fund* (USF) levies. VoIP as a bypass technology initially encountered some resistance in countries where incumbent service providers had much to lose to bypass operations. However, privatization of national telephone companies and a worldwide movement toward *unbundled local loop* (ULL) gives impetus to the adoption of VoIP and softswitch technologies as voice technologies contributing to an improved teledensity and its resulting improved economic infrastructure.

# Economic Advantage of Softswitch

Supposing one accepts the parity of a softswitch and a Class 4 or 5 switch in terms of scalability, reliability, QoS, signaling, and features, as argued previously, a softswitch has one more advantage: price.

A lower barrier to entry and exit encourages alternative service providers to enter the market. Some types of service providers that could be encouraged to offer voice services in competition to incumbent telephone service providers include *Internet service providers* (ISPs), cable TV companies, electric utility companies, *application service providers* (ASPs), municipalities, and wireless service providers.

# Disruptive or Deconstructive Technology?

In his landmark book, *The Innovator's Dilemma,* author Clayton Christensen describes how disruptive technologies have precipitated the failure of leading products, and the well-managed firms behind them. Christensen defines generic criteria to identify disruptive technologies regardless of market. Such technologies have the potential to replace mainstream technologies and remake markets. They are, says Christensen, "typically cheaper, simpler, smaller, and, frequently, more convenient" than their mainstream counterparts.

Relative to Class 4 and Class 5 switches, softswitch is disruptive technology. Although no vendor or ILEC has ever been driven out of business by softswitch, softswitch technologies are potentially disruptive to both incumbent telephone companies and switch vendors. It can also be argued that the telephone industry has been "deconstructed" by the Internet or Internet-related technologies. Instead of making long-distance calls or sending faxes over the PSTN, business people now send emails or use web sites. Long-distance calls may be placed over VoIP networks. Increasing alternatives decrease demand on the legacy telephone network and thereafter for telephone switching equipment.

# 2

# The Public Switched Telephone Network (PSTN)

An understanding of the workings of the *Public Switched Telephone Network* (PSTN) is best grasped by understanding its three major components: access, switching, and transport (see Figure 2-1). Each element has evolved over the 100-plus year history of the PSTN. Access pertains to how a user accesses the network. Switching refers to how a call is "switched" or routed through the network, and transport describes how a call travels or is "transported" over the network.

# Access

Access refers to how the user accesses the telephone network. For most users, access is gained to the network via a telephone handset. Transmission and reception is via diaphragms where the mouthpiece converts the air pressure of voice into an analog electromagnetic wave for transmission to the switch. The earpiece performs this process in reverse. The most sophisticated aspect of the handset is its *Dual-Tone Multifrequency* (DTMF) function, which signals the switch by tones. The handset is usually connected to the central office (where the switch is located) via copper wire known as *twisted pair* because, in most cases, it consists of a twisted pair of copper wire. The stretch of copper wire connects the telephone handset to the central office. Everything that runs between the subscriber and the central office is known as *outside plant*. Telephone equipment at the subscriber end is called *customer premise equipment* (CPE).

**Figure 2-1**
The three components of a telephone network: access, switching, and transport

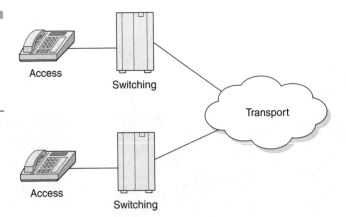

## Switching

The PSTN is a star network; that is, every subscriber is connected to another via at least one if not many hubs known as offices. In those offices are switches. Very simply, local offices are used for local service connections and tandem offices for long-distance service. Local offices, better known as central offices, use Class 5 switches, and tandem offices use Class 4 switches. Figure 2-2 details the relationship between Class 4 and 5 switches. A large city might have several central offices. Denver (population 2 million), for example, is estimated to have almost 40 central offices. Central offices in a large city often take up much of a city block and are recognizable as large brick buildings with no windows.

The first telephone switches were human. Taking a telephone handset off hook alerted a telephone operator of the caller's intention to place a call. The caller informed the operator of their intended called party and the operator set up the call by manually connecting the two parties.

Mechanical switching is credited to Almon Stowger, an undertaker in Kansas City, Missouri, who realized he was losing business when families of the deceased picked up their telephone handset and simply asked the operator to connect them with "the undertaker." The sole operator in this

**Figure 2-2**

The traditional relationship of Class 4, Class 5, and data networks

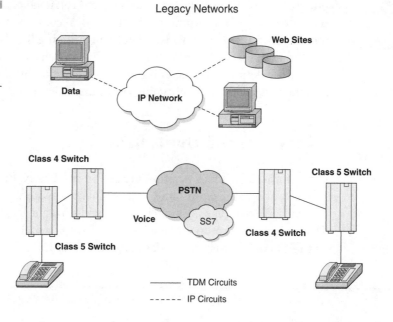

town was engaged to an undertaker competing with Stowger. This competing undertaker had promised to marry the operator once he had the financial means to do so. The operator, in turn, was more than willing to help him achieve that goal.

Stowger, realizing he was losing business to his competitor due to the intercession of the telephone operator, proceeded to invent an electromechanical telephone handset and switch that enabled the caller, by virtue of dialing the called party's number, to complete the connection without human intervention. Telephone companies realized the enormous savings in manpower (or womanpower as the majority of telephone operators at the time were women) by automating the call setup and takedown process. Stowger switches (also known as crossbar switches) can still be found in the central offices of rural America and lesser developed countries.

Stowger's design remained the predominant telephone switching technology until the mid-1970s. Beginning in the '70s, switching technology evolved to mainframe computers; that is, no moving parts were used and the computer telephony applications made such features as conferencing and call forwarding possible. In 1976, AT&T installed its first *#4 Electronic Switching System* (4ESS) tandem switch. This was followed shortly thereafter with the 5ESS as a central office switch. ESS central office switches did not require a physical connection between incoming and outgoing circuits. Paths between the circuits consisted of temporary memory locations that enabled the temporary storage of traffic. For an ESS system, a computer controls the assignment, storage, and retrieval of memory locations so that a portion of an incoming line (time slot) could be stored in temporary memory and retrieved for insertion to an outgoing line. This is called a *time slot interchange* (TSI) memory matrix. The switch control system maps specific time slots on an incoming communication line (such as a DS3) to specific time slots on an outgoing communication line.[1]

## Class 4 and 5 Switching

Class 4 and 5 switches are the "brains" of the PSTN. Figure 2-3 illustrates the flow of a call from a handset to a Class 5 switch, which in turn hands the call off to a Class 4 switch for routing over a long-distance network. That call may be routed through other Class 4 switches before terminating at the Class 5 switch at the destination end of the call before being passed

---

[1]Harte, Lawrence. *Telecom Made Simple*. Fuquay-Varina, NC: APDG Publishing, 2002.

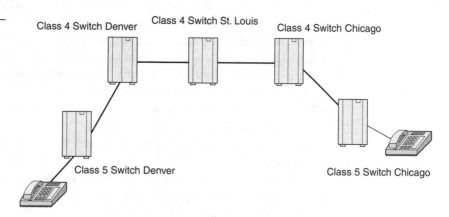

**Figure 2-3**
Relationship of Class
4 and 5 switching

on to the terminating handset. Class 5 switches handle local calling and Class 4 switches handle long-distance calls. The performance metrics for the Class 4 and 5 have been reliability, scalability, *quality of service* (QoS), signaling, and features.

**Class 4 and 5 Architecture** One reason for the reputation of Class 4 and 5 switches being reliable is that they have been tested by time in the legacy market. Incremental improvements to the 4ESS included new interfaces, hardware, software, and databases to improve *Operations, Administration, Maintenance, and Provisioning* (OAM&P). The inclusion of the 1A processor improved memory in the 4 and 5ESS mainframe, allowing for translation databases. Ultimately, those databases were interfaced with the *Centralized Automatic Reporting on Trunks* (CAROT). Later, integrated circuit chips replaced the magnetic core stores and improved memory and boosted the *Busy Hour Call Attempt* (BHCA) capacity to 700,000 BHCAs.[2]

**Class 4 and 5 Components** The architecture of the Class 4 and 5 switch is the product of 25-plus years of design evolution. For the purposes of this discussion, the Nortel DMS-250, one of the most prevalent products in the North American Class 4 market, is used as a real-world example. The other

[2]Chapuis, Robert, and Amos Joel. "In the United States, AT&T's Digital Switch Entry No. 4 ESS, First Generation Time Division Digital Switch." *Electronics, Computers, and Telephone Systems.* New York: North Holland Publishing, 1990, p. 337–338.

leading product in this market is the 4ESS from Lucent Technologies. For local offices or Class 5, the most prevalent product is the 5ESS from Lucent. DMS-250 hardware, for example, is redundant for reliability and decreased downtime during upgrades. It has a reliability rating of 99.999 percent (the five 9s), which meets the industry metric for reliability. The modular design of the hardware enables the system to scale from 480 to over 100,000 DS0s (individual phone lines). The density, or number of phone lines the switch can handle, is one metric of scalability. The DMS-250 is rated at 800,000 BHCAs. Tracking BHCAs on a switch is a measure of call-processing capability and is another metric for scalability.

Key hardware components of the DMS-250 system include the DMS core, switch matrix, and trunk interface. The DMS core is the *central processing unit* (CPU) and memory of the system, handling high-level call processing, system control functions, system maintenance, and the installation of new switch software.

The DMS-250 switching matrix switches calls to their destinations. Its nonblocking architecture enables the switch to communicate with peripherals through fiber optic connections. The trunk interfaces are peripheral modules that form a bridge between the DMS-250 switching matrix and the trunks it serves. They handle voice and data traffic to and from customers and other switching systems. DMS-250 trunk interfaces terminate DS-1, *Integrated Services Digital Network* (ISDN) *Primary Rate Interface* (PRI), X.75/X.75 packet networking, and analog trunks. They also accommodate test and service circuits used in office and facility maintenance. It is important to note that the Class 4 switching matrix is a part of the centralized architecture of the Class 4. Unlike the media gateways in a softswitch solution, it must be collocated with the other components of the Class 4.

DMS-250 billing requires the maintenance of real-time, transaction-based billing records for many thousands of customers and scores of variants in service pricing. The DMS-250 system automatically provides detailed data, formats the data into call detail records, and constructs bills.[3]

## Private Branch Exchange (PBX)

As the name would imply, a *private branch exchange* (PBX) is a switch owned and maintained by a business with many (20 or more) users. A key

---

[3]Nortel Networks. "Product Service Information-DMS300/250 System Advantage." www. nortel.com, 2001.

system is used by smaller offices. PBXs and key systems today are computer based and enable soft changes to be made through an administration terminal or PC. Unless the business has a need for technical telecommunications personnel on staff for other reasons, the business will normally contract with their vendor for routine adds, moves, and changes of telephone equipment.

PBX systems are often equipped with key assemblies and systems, including voice mail, call accounting, a local maintenance terminal, and a dial-in modem. The voice mail system is controlled by the PBX and only receives calls when the PBX software determines a message can be left or retrieved. The call accounting system receives system message details on all call activities that occur within the PBX. The local terminal provides onsite access to the PBX for maintenance activities. The dial-in capability also provides access to the PBX for maintenance activities.[4]

## Centrex

After PBXs caught on in the industry, local exchange carriers began to lose some of their more lucrative business margins. The response to the PBX was Centrex. Centrex is a service offered by a local telephone service provider (primarily to businesses) that enables the customer to have features that are typically associated with a PBX. These features include three- or four-digit dialing, intercom features, distinctive line ringing for inside and outside lines, voice mail, call-waiting indication, and others. Centrex services flourished and still have a place for many large, dispersed entities such as large universities and major medical centers.

One of the major selling points for Centrex is the lack of capital expenditure up front. That, coupled with the reliability associated with Centrex due to its location in the telephone company central office, has kept Centrex as the primary telephone system in many of the businesses referenced previously. PBXs, however, have cut into what was once a lucrative market for the telephone companies and are now the rule rather than the exception for business telephone service. This has come about because of inventive ways of funding the initial capital outlay and the significantly lower operating cost of a PBX versus a comparable Centrex offering.

---

[4]Harte, Lawrence. *Telecom Made Simple*. Fuquay-Varina, NC: APDG Publishing, 2002.

# Multiplexing

The earliest approach to getting multiple conversations over one circuit was *frequency division multiplexing* (FDM). FDM was made possible by the vacuum tube where the range of frequencies was divided into parcels that were distributed among subscribers. In the first FDM architectures, the overall system bandwidth was 96 kHz. This 96 kHz could be divided among a number of subscribers into, for example, 5 kHz per subscriber, meaning almost 20 subscribers could use this circuit.

FDM is an analog technology and suffers from a number of shortcomings. It is susceptible to picking up noise along the transmission path. This FDM signal loses its power over the length of the transmission path. FDM requires amplifiers to strengthen the signal over that path. However, the amplifiers cannot separate the noise from the signal and the end result is an amplified noisy signal.

The improvement over FDM was *time division multiplexing* (TDM). TDM was made possible by the transistor that arrived in the market in the 1950s and 1960s. As the name would imply, TDM divides the *time* rather than the frequency of a signal over a given circuit. Although FDM was typified by "some of the frequency all of the time," TDM is "all of the frequency some of the time." TDM is a digital transmission scheme that uses a small number of discrete signal states. Digital carrier systems have only three valid signal values: one positive, one negative, and zero. Everything else is registered as noise. A repeater, known as a regenerator, can receive a weak and noisy digital signal, remove the noise, reconstruct the original signal, and amplify it before transmitting the signal onto the next segment of the transmission facility. Digitization brings with it the advantages of better maintenance and troubleshooting capability, resulting in better reliability. Also, a digital system enables improved configuration flexibility.

TDM has made the multiplexer, also known as the channel bank, possible. In the United States, the multiplexer or "mux" enables 24 channels per single four-wire facility. This is called a T-1, DS1, or T-Carrier. Outside North America and Japan, it is 32 channels per facility and known as E1. These systems came on the market in the early 1960s as a means to transport multiple channels of voice over expensive transmission facilities.

# Voice Digitization via Pulse Code Modulation

One of the first processes in the transmission of a telephone call is the conversion of an analog signal into a digital one. This process is called *pulse code modulation* (PCM). This is a four-step process consisting of *pulse amplitude modulation* (PAM) sampling, companding, quantization, and encoding.

**Pulse Amplitude Modulation (PAM)**   The first stage in PCM is known as PAM. In order for an analog signal to be represented as a digitally encoded bitstream, the analog signal must be sampled at a rate that is equal to twice the bandwidth of the channel over which the signal is to be transmitted. As each analog voice channel is allocated 4 kHz of bandwidth, each voice signal is sampled at twice that rate, or 8,000 samples per second. In a T-Carrier, the standard in North America and Japan, each channel is sampled every one eight-thousandth of a second in rotation, resulting in the generation of 8,000 pulse amplitude samples from each channel every second. If the sampling rate is too high, too much information is transmitted and bandwidth is wasted. If the sampling rate is too low, aliasing may result. Aliasing is the interpretation of the sample points as a false waveform due to the lack of samples.

**Companding**   The second process of PCM is companding. Companding is the process of compressing the values of the PAM samples to fit the non-linear quantizing scale that results in bandwidth savings of more than 30 percent. It is called companding as the sample is compressed for transmission and expanded for reception.[5]

**Quantization**   The third stage in PCM is quantization. In quantization, values are assigned to each sample within a constrained range. In using a limited number of bits to represent each sample, the signal is quantized. The difference between the actual level of the input analog signal and the digitized representation is known as quantization noise. Noise is a detraction to voice quality and it is necessary to minimize noise. The way to do this is to use more bits, thus providing better granularity. In this case, an inevitable trade-off takes place bewteen bandwidth and quality. More bandwidth usually improves signal quality, but bandwidth costs money. Service providers, whether using TDM or *Voice over IP* (VoIP) for voice transmission will always have to choose between quality and bandwidth. A process known as nonuniform quantization involves the usage of smaller

---

[5]Shepard, Steven. *SONET/SDH Demystified.* New York: McGraw-Hill, 2001. p. 15–21.

quantization steps at smaller signal levels and larger quantization steps for larger signal levels. This gives the signal greater granularity or quality at low signal levels and less granularity (quality) at high signal levels. The result is to spread the signal-to-noise ratio more evenly across the range of different signals and to enable fewer bits to be used compared to uniform quantization. This process results in less bandwidth being consumed than for uniform quantization.[6]

**Encoding**   The fourth and final process in PCM is encoding the signal. This is performed by a *codec* (coder/decoder). Three types of codecs exist: waveform codecs, source codecs (also known as vocoders), and hybrid codecs.   Waveform codecs sample and code an incoming analog signal without regard to how the signal was generated. Quantized values of the samples are then transmitted to the destination where the original signal is reconstructed, at least to a certain approximation of the original. Waveform codecs are known for simplicity with high-quality output. The disadvantage of waveform codecs is that they consume considerably more bandwidth than the other codecs. When waveform codecs are used at low bandwidth, speech quality degrades markedly.

Source codecs match an incoming signal to a mathematical model of how speech is produced. They use the linear predictive filter model of the vocal tract, with a voiced/unvoiced flag to represent the excitation that is applied to the filter. The filter represents the vocal tract and the voice/unvoiced flag represents whether a voiced or unvoiced input is received from the vocal chords. The information transmitted is a set of model parameters as opposed to the signal itself. The receiver, using the same modeling technique in reverse, reconstructs the values received into an analog signal.

Source codecs also operate at low bit rates and reproduce a synthetically sounding voice. Using higher bit rates does not result in improved voice quality. Vocoders (source codecs) are most widely used in private and military applications.

Hybrid codecs are deployed in an attempt to derive the benefits from both technologies. They perform some degree of waveform matching while mimicking the architecture of human speech. Hybrid codecs provide better voice quality at low bandwidth than waveform codecs. Table 2-1 provides an outline of the different ITU codec standards and Table 2-2 lists the parameters of the voice codecs.

---

[6]Collins, Daniel. *Carrier Grade Voice Over IP.* New York: McGraw-Hill, 2001. p. 95–96.

**Table 2-1**

Descriptions of
voice codecs (ITU)

| ITU Standard | Description |
|---|---|
| P.800 | A subjective rating system to determine the *Mean Opinion Score* (MOS) or the quality of telephone connections |
| G.114 | A maximum one-way delay end to end for a VoIP call (150 ms) |
| G.165 | Echo cancellers |
| G.168 | Digital network echo cancellers |
| G.711 | PCM of voice frequencies |
| G.722 | 7 kHz audio coding within 64 Kbps |
| G.723.1 | A dual-rate speech coder for multimedia communications transmitting at 5.3 and 6.3 Kbps |
| G.729 | Coding for speech at 8 Kbps using *conjugate-structure algebraic code-excited linear-prediction* (CS-ACELP) |
| G.729A | Annex A reduced complexity 8 Kbps CS-ACELP speech codec |
| H.323 | A packet-based multimedia communications system |
| P.861 | Specifies a model to map actual audio signals to their representations inside the human head |
| Q.931 | Digital subscriber signaling system number 1 ISDN user-network interface layer 3 specification for basic call control |

**Table 2-2**

Parameters of voice
codecs

| Standard | Data rate (Kbps) | Delay (ms) | MOS | Codec |
|---|---|---|---|---|
| G.711 | 64 | 0.125 | 4.8 | Waveform |
| G.721, G.723, G.726 | 16,24,32,40 | 0.125 | 4.2 | |
| G.728 | 16 | 2.5 | 4.2 | |
| G.729 | 8 | 10 | 4.2 | |
| G.723.1 | 5.3, 6.3 | 30 | 3.5, 3.98 | |

**Popular Speech Codecs**  Codecs are best known for the sophisticated compression algorithms they introduce into a conversation. Bandwidth costs service providers money. The challenge for many service providers is to squeeze as much traffic as possible into one "pipe," that is one channel. Most codecs allow multiple conversations to be carried on one 64 kbps channel. There is an inevitable trade off in compression for voice quality in the conversation. The challenge for service providers is to balance the economics of compression with savings in bandwidth costs.

**G.711**   G.711 is the best-known coding technique in use today. It is a waveform codec and is the coding technique used in circuit-switched telephone networks all over the world. G.711 has a sampling rate of 8,000 Hz. If uniform quantization were to be used, the signal levels commonly found in speech would be such that at least 12 bits per sample would be needed, giving it a bit rate of 96 Kbps. Nonuniform quantization is used with eight bits used to represent each sample. This quantization leads to the well-known 64 Kbps DS0 rate. G.711 is often referred to as PCM. G.711 has two variants: A-law and mu-law. Mu-law is used in North America and Japan where T-Carrier systems prevail. A-law is used everywhere else in the world. The difference between the two is the way nonuniform quantization is performed. Both are symmetrical at approximately zero. Both A-law and mu-law offer good voice quality with a MOS of 4.3, with 5 being the best and 1 being the worst. Despite being the predominant codec in the industry, G.711 suffers one significant drawback; it consumes 64 Kbps in bandwidth. Carriers seek to deliver voice quality using little bandwidth, thus saving on operating costs.

**G.728 LD-CELP**   *Code-Excited Linear Predictor* (LD-CELP) codecs implement a filter and contain a codebook of acoustic vectors. Each vector contains a set of elements in which the elements represent various characteristics of the excitation signal. CELP coders transmit to the receiving end a set of information determining filter coefficients, gain, and a pointer to the chosen excitation vector. The receiving end contains the same code book and filter capabilities so that it reconstructs the original signal.   G.728 is a backward-adaptive coder as it uses previous speech samples to determine the applicable filter coefficients. G.728 operates on five samples at one time. That is, 5 samples at 8,000 Hz are needed to determine a codebook vector and filter coefficients based upon previous and current samples. Given a coder operating on five samples at a time, a delay of less than 1 millisecond is the result. Low delay equals better voice quality.

The G.728 codebook contains 1,024 vectors, which requires a 10-bit index value for transmission. It also uses 5 samples at a time taken at a rate of 8,000 per second. For each of those 5 samples, G.728 results in a transmitted bit rate of 16 Kbps. Hence, G.728 has a transmitted bit rate of 16 Kbps. Another advantage here is that this coder introduces a delay of 0.625 milliseconds with an MOS of 3.9. The difference from G.711's MOS of 4.3 is imperceptible to the human ear. The bandwidth savings between G.728's 16 Kbps per conversation and G.711's 64 Kbps per conversation make G.728 very attractive to carriers given the savings in bandwidth.

**G.723.1 ACELP**  G.723.1 ACELP can operate at either 6.3 Kbps or 5.3 Kbps with the 6.3 Kbps providing higher voice quality. Bit rates are contained in the coder and decoder, and the transition between the two can be made during a conversation. The coder takes a bank-limited input speech signal that is sampled a 8,000 Hz and undergoes uniform PCM quantization, resulting in a 16-bit PCM signal. The encoder then operates on blocks or frames of 240 samples at a time. Each frame corresponds to 30 milliseconds of speech, which means that the coder causes a delay of 30 milliseconds. With a look-ahead delay of 7.5 milliseconds, the total algorithmic delay is 37.5 milliseconds. G.723.1 gives an MOS of 3.8, which is highly advantageous in regards to the bandwidth used. The delay of 37.5 milliseconds one way does present an impediment to good quality, but the round-trip delay over varying aspects of a network determines the final delay and not necessarily the codec used.

**G.729**  G.729 is a speech coder that operates at 8 Kbps. This coder uses input frames of 10 milliseconds, corresponding to 80 samples at a sampling rate of 8,000 Hz. This coder includes a 5-millisecond look-ahead, resulting in an algorithmic delay of 15 milliseconds (considerably better than G.723.1). G.729 uses an 80-bit frame. The transmitted bit rate is 8 Kbps. Given that it turns in an MOS of 4.0, G.729 is perhaps the best trade-off in bandwidth for voice quality.   The previous paragraphs provide an overview of the multiple means of maximizing the efficiency of transport via the PSTN. We find today that TDM is almost synonymous with circuit switching. Telecommunications engineers use the term TDM to describe a circuit-switched solution. A 64 Kbps G.711 codec is the standard in use on the PSTN. The codecs described in the previous pages apply to VoIP as well. VoIP engineers seeking to squeeze more conversations over valuable bandwidth have found these codecs very valuable in compressing VoIP conversations over an IP circuit.[7]

# Signaling

Signaling describes the process of how calls are set up and torn down. Generally speaking, there are three main functions of signaling: supervision, alerting, and addressing. Supervision refers to monitoring the status of a line or circuit to determine if there is traffic on the line. Alerting deals with the ringing of a phone indicating the arrival of an incoming call. Addressing is the routing of a call over a network. As telephone networks matured,

---

[7]Ibid.

individual nations developed their proprietary signaling systems. Ultimately, there become a signaling protocol for every national phone service in the world. Frankly, it is a miracle that international calls are ever completed given the complexity of interfacing national signaling protocols.

**Signaling System 7 (SS7)**    For much of the history of circuit-switched networks, signaling followed the same path as the conversation. This is called *Channel-Associated Signaling* (CAS) and is still in wide use today. R1 *Multifrequency* (MF) used in North American markets and *R2 Multi-Frequency Compelled* (RFC) used elsewhere in the world are the best examples of this. Another name for this is in-channel signaling.   The newer technology for signaling is called *Common Channel Signaling* (CCS), also known as out-of-band signaling. CCS uses a separate transmission path for call signaling and not the bearer path for the call. This separation enables the signaling to be handled in a different manner to the call. This enables signaling to be managed by a network independent of the transport network. Figure 2-4 details the difference between CAS and CCS.

**Figure 2-4**
CAS and CCS

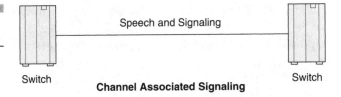

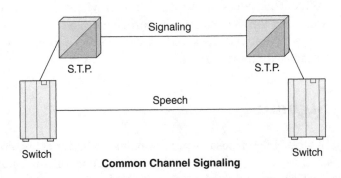

---

[8]Stallings, William. *ISDN and Broadband ISDN with Frame Relay and ATM.* New York: Prentice Hall, 1995. p.292.

*Signaling System 7* (SS7) is the standard for CCS with many national variants throughout the world (such as Mexico's NOM-112). It routes control messages through the network to perform call management (setup, maintenance, and termination) and network management functions. Although the network being controlled is circuit switched, the control signaling is implemented using packet-switching technology. In effect, a packet-switched network is overlaid on a circuit-switched network in order to operate and control the circuit-switched network. SS7 defines the functions that are performed in the packet-switched network but does not dictate any particular hardware implementation.[8]

The SS7 network and protocol are used for the following:

- Basic call setup, management, and tear down
- Wireless services such as *personal communications services* (PCS), wireless roaming, and mobile subscriber authentication
- *Local number portability* (LNP)
- Toll-free (800/888) and toll (900) wireline services
- Enhanced call features such as call forwarding, calling party name/number display, and three-way calling
- Efficient and secure worldwide telecommunications

**Signaling Links**  SS7 messages are exchanged between network elements over 56 or 64 Kbps bidirectional channels called signaling links. Signaling occurs out of band on dedicated channels rather than in-band on voice channels. Compared to in-band signaling, out-of-band signaling provides faster call setup times (compared to in-band signaling using MF signaling tones), more efficient use of voice circuits, support for *Intelligent Network* (IN) services that require signaling to network elements without voice trunks (such as database systems), and improved control over fraudulent network usage.

**Signaling Points**  Each signaling point in the SS7 network is uniquely identified by a numeric point code. Point codes are carried in signaling messages exchanged between signaling points to identify the source and destination of each message. Each signaling point uses a routing table to select the appropriate signaling path for each message. Three kinds of signaling points are used in the SS7 network: *service switching points* (SSP), *signal transfer points* (STP), and *service control points* (SCP), as shown in Figure 2-5.

SSPs are switches that originate, terminate, or tandem calls. An SSP sends signaling messages to other SSPs to set up, manage, and release voice

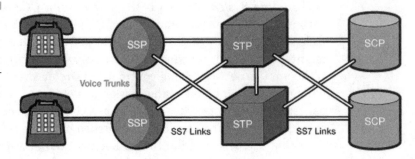

circuits required to complete a call. An SSP may also send a query message to a centralized database (an SCP) to determine how to route a call (such as a toll-free 1-800/888 call in North America). An SCP sends a response to the originating SSP containing the routing number(s) associated with the dialed number. An alternate routing number may be used by the SSP if the primary number is busy or the call is unanswered within a specified time. Actual call features vary from network to network and from service to service.

Network traffic between signaling points may be routed via a packet switch called an STP. An STP routes each incoming message to an outgoing signaling link based on routing information contained in the SS7 message. Because it acts as a network hub, an STP provides improved utilization of the SS7 network by eliminating the need for direct links between signaling points. An STP may perform global title translation, a procedure by which the destination signaling point is determined from digits present in the signaling message (such as the dialed 800 number, the calling card number, or mobile subscriber identification number). An STP can also act as a firewall to screen SS7 messages exchanged with other networks.

Because the SS7 network is critical to call processing, SCPs and STPs are usually deployed in mated-pair configurations in separate physical locations to ensure network-wide service in the event of an isolated failure. Links between signaling points are also provisioned in pairs. Traffic is shared across all links in the linkset. If one of the links fails, the signaling traffic is rerouted over another link in the linkset. The SS7 protocol provides both error correction and retransmission capabilities to enable continued service in the event of signaling point or link failures.

**SS7 Signaling Link Types** Signaling links are logically organized by link type (A through F) according to their use in the SS7 signaling network (see Figure 2-6 and Table 2-3).

**Figure 2-6**

SS7 signaling link types (Source: Performance Technologies)

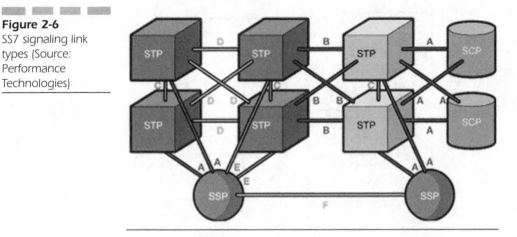

| **Table 2-3** | A link | An A (access) link connects a signaling end point (an SCP or SSP) to an STP. Only messages originating from or destined to the signaling end point are transmitted on an A link. |
|---|---|---|
| Descriptions of SS7 links | B link | B (bridge) links connect an STP to another STP. Typically, a quad of B links interconnect peer (or primary) STPs (the STPs from one network to the STPs of another). The distinction between a B link and a D link is rather arbitrary. For this reason, such links may be referred to as B/D links. |
| | C link | C (cross) links connect STPs performing identical functions into a mated pair. They are used only when an STP has no other route available to a destination signaling point due to link failure(s). Note that SCPs may also be deployed in pairs to improve reliability. Unlike STPs, however, signaling links do not interconnect mated SCPs. |
| | D link | D (diagonal) links connect a secondary (local or regional) STP pair to a primary (internetwork gateway) STP pair in a quad-link configuration. Secondary STPs within the same network are connected via a quad of D links. The distinction between a B link and a D link is rather arbitrary. For this reason, such links may be referred to as B/D links. |
| | E link | An E (extended) link connects an SSP to an alternate STP. E links provide an alternate signaling path if an SSP's home STP cannot be reached via an A link. E links are not usually provisioned unless the benefit of a marginally higher degree of reliability justifies the added expense. |
| | F link | An F (fully associated) link connects two signaling end points (SSPs and SCPs). F links are not usually used in networks with STPs. In networks without STPs, F links directly connect signaling points. |

Source: Performance Technologies

**Figure 2-7**
The OSI Reference
Model and the SS7
protocol stack
(Source: Performance
Technologies)

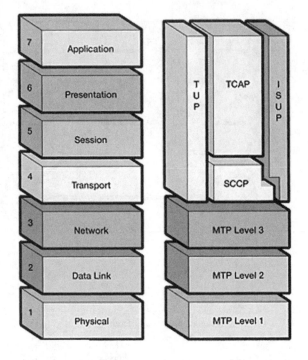

**SS7 Protocol Stack**   The hardware and software functions of the SS7 protocol are divided into functional abstractions called levels. These levels map loosely to the *Open Systems Interconnect* (OSI) seven-layer model defined by the *International Standards Organization* (ISO), as shown in Figure 2-7.

**Message Transfer Part**   The *Message Transfer Part* (MTP) is divided into three levels. The lowest level, MTP level 1, is equivalent to the OSI physical layer. MTP level 1 defines the physical, electrical, and functional characteristics of the digital signaling link. Physical interfaces defined include E-1 (2,048 Kbps; 32 64-Kbps channels), DS-1 (1,544 Kbps; 24 64-Kbps channels), V.35 (64 Kbps), DS-0 (64 Kbps), and DS-0A (56 Kbps).   MTP level 2 ensures accurate end-to-end transmission of a message across a signaling link. Level 2 implements flow control, message sequence validation, and error checking. When an error occurs on a signaling link, the message (or set of messages) is retransmitted. MTP level 2 is equivalent to the OSI data link layer.

MTP level 3 provides message routing between signaling points in the SS7 network. MTP level 3 reroutes traffic away from failed links and signaling points, and it controls traffic when congestion occurs. MTP level 3 is equivalent to the OSI network layer.

**ISDN User Part (ISUP)**   The *ISDN User Part* (ISUP) defines the protocol used to set up, manage, and release trunk circuits that carry voice and data between terminating line exchanges (between a calling party and a called party). ISUP is used for both ISDN and non-ISDN calls. However, calls that originate and terminate at the same switch do not use ISUP signaling.

**Telephone User Part (TUP)**   In some parts of the world (such as China and Brazil), the *Telephone User Part* (TUP) is used to support basic call setup and teardown. TUP handles analog circuits only. In many countries, ISUP has replaced TUP for call management.

**Signaling Connection Control Part (SCCP)**   SCCP provides connectionless and connection-oriented network services and *global title translation* (GTT) capabilities above MTP level 3. A global title is an address (a dialed 800 number, calling card number, or mobile subscriber identification number) that is translated by SCCP into a destination point code and subsystem number. A subsystem number uniquely identifies an application at the destination signaling point. SCCP is used as the transport layer for TCAP-based services.

**Transaction Capabilities Applications Part (TCAP)**   TCAP supports the exchange of noncircuit-related data between applications across the SS7 network using the SCCP connectionless service. Queries and responses sent between SSPs and SCPs are carried in TCAP messages. For example, an SSP sends a TCAP query to determine the routing number associated with a dialed 800/888 number and to check the *personal identification number* (PIN) of a calling card user. In mobile networks (IS-41 and GSM), TCAP carries *Mobile Application Part* (MAP) messages sent between mobile switches and databases to support user authentication, equipment identification, and roaming.

# The Advanced Intelligent Network (AIN)

How are features delivered? In one concept, features are made possible by the *Advanced Intelligent Network* (AIN) and SS7. The AIN is a telephone network architecture that separates service logic from switching equipment, enabling new services to be added without having to redesign switches to support new services. It encourages competition among service providers as it makes it easier for a provider to add services, and it offers customers more service choices. Developed by Bell Communications Research, AIN is recognized as an industry standard in North America.

The AIN was a concept promoted by large telephone companies throughout the 1980s to promote their architecture for the 1990s and beyond. Two consistent themes characterize the AIN. One is that the network can control the routing of calls within it from moment to moment based on some criteria other than that of finding a path through the network for the call based on the dialed number. The other is that the originator or receiver of the call can inject intelligence into the network and affect the flow of the call. That intelligence is provided through the use of databases in a network.

The foundation of the AIN architecture is SS7 (see Figure 2-8). SS7 enables a wide range of services to be provided to the end-user. An SCP is a network entity that contains additional logic and that can be used to offer advanced services. To use the service logic of the SCP, a switch needs to contain functionality that will enable it to act upon instructions from the SCP. In such a case, the switch is known as an SSP. If a particular service needs to be invoked, the SSP sends a message to the SCP asking for instruction. The SCP, based upon data and service logic that is available, will tell the SSP which actions need to be taken.[9]

How does AIN work? A telephone caller dials a number that is received by a switch at the telephone company central office. The switch, also known as the signaling point, forwards the call over an SS7 network to an SCP where the service logic is located. The SCP identifies the service requested from part of the number that was dialed and returns information about how to handle the call to the signaling point. Examples of services that the SCP might provide include area number calling services, disaster recovery services, do not disturb services, and 800 toll-free and 5-digit extension dialing services.

**Figure 2-8**
AIN release 1
architecture
(Source: Telcordia)

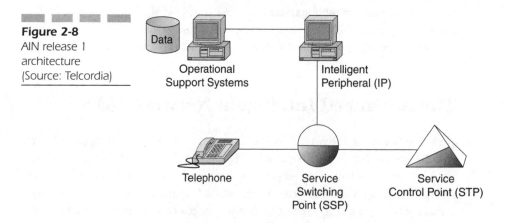

Data

Operational
Support Systems

Intelligent
Peripheral (IP)

Telephone

Service
Switching
Point (SSP)

Service
Control Point (STP)

---

[9]Collins, Michael. *Carrier Grade Voice Over IP.* New York: McGraw-Hill, 2001. p. 311.

In some cases, the call can be handled more quickly by an *intelligent peripheral* attached to the SSP over a high-speed connection. For example, a customized voice announcement can be delivered in response to the dialed number or a voice call can be analyzed and recognized. In addition, an adjunct facility can be added directly to the SSP for a high-speed connection to additional, undefined services.[10]

SCPs have two complementary tasks. First, they host the application functionality on which service logic is installed after services are created. Secondly, the SCP controls functionalities developed by SCP vendors. The SCP contains programmable, service-independent capabilities (or service logic) that are under the control of service providers. As a separate offering, the SCP can contain service-specific data that can be customized by either service providers or their customers (at the service provider's discretion). In addition to its programmable functionality, the SCP provides SS7 interface to switching systems.

A third element in the AIN architecture is the intelligent peripheral. Intelligent peripherals provide resources such as voice announcements, voice recognition, and DTMF digit collection. Intelligent peripherals support flexible interaction between the end user and the network.

The two main benefits of AIN are its capabilities to improve existing services and to develop new services as sources of revenue. To meet these goals, service providers must introduce new services rapidly, provide service customization, establish vendor independence, and create open interfaces. AIN technology uses an embedded base of stored program-controlled switching systems and the SS7 network. At the same time, AIN technology enables the separation of service-specific functions and data from other network resources. This feature reduces the dependency on switching system vendors for software developments and delivery schedules. In theory, service providers have more freedom to create and customize services.[11]

## Features

*Custom Local Area Signaling Service* (CLASS) features are basic services available in each *local access and transport area* (LATA). The features and

---

[10]Search Networking. "Advanced Intelligent Networks," p.1, http://searchnetworking.techtarget.com. (This URL is no longer valid.)

[11]Telcordia Technologies. "Intelligent Networks (IN)." A white paper hosted by the International Engineering Consortium at www.iec.org.

the services they enable are a function of Class 5 switches and SS7 networks. The Class 4 switch offers no features of its own. It transmits the features of Class 5. With almost three decades of development, the Class 4 switch has a well-established history of interoperability with the features offered by the Class 5 and SS7 networks. Features often enable service provider systems to generate high margins that, of course, equate to stronger revenue streams.

Examples of features offered through the DMS-250 system can be grouped into two major categories: basic and enhanced services. The basic services include 1+, 800/900 service, travel cards, account codes, PIN numbers, operator access, speed dialing, hotline service, *automatic number identification* (ANI) screening, *virtual private networks* (VPNs), calling cards, and call detail recording. Enhanced services include information database services (NXX number services, authorization codes, calling card authorization, and debit/prepaid card services), and routing and screening (includes Carrier Identification Code [CIC] routing, time of day screening, ANI screening, and class of service screening). Enhanced features also include enterprise networks, data and video services (dedicated access lines, ISDN PRI services, dialable wideband services, and switched 56 Kpbs), and multiple dialing plans (full 10-digit routing, 7-digit VPN routing, 15-digit international dialing, speed dialing, and hotline dialing). Most of these features have been standards on the DMS-250 and other Class 4 switches for many years.

This long list of features is evidence of the importance of features in the legacy market in which they were developed. Service providers are reluctant to give up these features and the higher margins they generate. In the converging market, features are equally important to reliability because service providers don't want to offer fewer features to their customers and they will want to continue to offer high-margin features.

# Performance Metrics for Class 4 and 5 Switches

To date, the basis for choosing a Class 4 or 5 switch architecture over that of softswitch has been reliability, scalability, good QoS, and well-known features and applications. The question now becomes, what happens when competing technologies meet or exceed the standards set around the legacy of the Class 4 or 5 switch?

**Reliability**   Not all businesses need to provide or rely on round-the-clock availability. For those that do, Table 2-4 from a 1998 Gartner Group study[12] illustrates how costly downtime can be within and across industries.

The telecommunications industry is a little different. If a Class 4 switch with 100,000 DS0s charging $0.05 per minute were to be down 1 hour, the service provider would lose $300,000 in revenue. Downtime is lost revenue. Five 9s of reliability is the standard for the legacy market. Service providers know their customers expect the same levels of reliability in any new market as they did in the legacy market. The focus of debate on this issue in the industry revolves around the question of how many nines of reliability a product can deliver. The PSTN or legacy voice network claims five 9s. What is meant by five 9s?

**One to Five 9s**   Availability is often expressed numerically as a percentage of uninterrupted productive time containing from one to five 9s. For instance, 99 percent availability, or two nines, equates to a certain amount of availability and downtime, as does 99.9 percent (three nines), and so on. Table 2-5 offers calculations for each of the five 9s based on 24-hour, year-round operation.

**How Does a Switch Achieve Five 9s?**   Five 9s are not the result of divine guidance given to Bellcore in the 1970s, but rather a process of engineering resulting in a high level of reliability. Reliability is enhanced when each component is replicated in a system. This is called redundancy. If one unit

**Table 2-4**

Costs of down time

| Industry | Application | Avg. cost per hour of downtime |
|---|---|---|
| Financial | Brokerage operations | $6,500,000 |
| Financial | Credit card sales | $2,600,000 |
| Media | Pay per view | $1,150,000 |
| Retail | Home shopping (TV) | $113,000 |
| Retail | Catalog sales | $90,000 |

[12]Nelson, Gene. "Architecting and Deploying High Availability Solutions: Business Drivers and Key Considerations." Compaq white paper. October 1998. Available online at ftp://ftp.compaq.com/pub/supportinformation/papers/ecg0641198.pdf.

**Table 2-5**

Availability and downtime: How the five 9s are calculated

| Availability | Downtime |
|---|---|
| 90% (one 9) | 36.5 days per year |
| 99% (two 9s) | 3.65 days per year |
| 99.9% (three 9s) | 8.76 hours per year |
| 99.99% (four 9s) | 52.55 minutes per year |
| 99.999% (five 9s) | 5.25 minutes per year |

fails, its replicated unit takes over. Redundancy is usually expressed in terms of a ratio of 1:1 where one replicated unit exists for every primary unit or N11 redundancy where there are N(N>1) replicated units per primary unit. Another mechanism to enhance reliability is to ensure no *single point of failure* (SPOF) exists on the system. That is, every mechanism has a backup in the event of the failure of one unit. Hot standby refers to having a replicated unit take over the functions of its primary unit for either planned or unplanned outages. Continuously available (reliable) systems rely exclusively on active replications to achieve transparency in masking both planned and unplanned outages.[13]

**Network Equipment Building Standards (NEBS)**   In addition to the five 9s, the other buzzword for reliability in the Class 4 market is *Network Equipment Building Standards* (NEBS). NEBS address the physical reliability of a switch. NEBS parameters are contained in Telcordia specification SR 3580, which contains the requirements for performance, quality, safety, and environmental metrics applicable to network equipment installed in a carrier's central office. Most North American carriers require equipment in their central offices or switching locations to be NEBS compliant. Tests include electrical safety, immunity from electromagnetic emissions, lightning and power faulting, and bonding and grounding evaluations. Between five 9s and NEBs, the Class 4 and 5 vendors have developed, through decades of experience, a reliable product that delivers superb uptime, and rapid recovery capabilities. As a result, service providers are often reluctant to experiment with new technologies.

---

[13]"The Five 9's Pilgrimage: Toward the Next Generation of High Availability Systems." A white paper from Clustra at www.clustra.com.

*Scalability* For financial reasons, a Class 4 or 5 switch must offer flexibility in scalability. Ideally, a service provider starts out with a chassis and a minimum capability in terms of DS0s (one DS0 is one phone line) and then adds more capacity as demand increases. This is considered a scalable solution and is preferable to a solution that requires either another chassis or a whole new system (known as a forklift upgrade) when demand grows beyond the initial installation. Nortel's DMS-250 Class 4 switch, for example, scales up to approximately 100,000 DS0s. The other metric for scalability is the call-processing power of the switch. The DMS-250, for example, can process 800,000 BHCAs.[14]

Another financial reason for scalability is to achieve a low cost per DS0 in purchases and operations. By buying a large switch with many DS0s, a service provider can negotiate a lower price per DS0 and improve the odds of being profitable in that market. For these financial reasons, scalability is important to service providers in both the legacy and converging markets. In terms of BHCAs, a Class 4, such as DMS-250, offers call-processing power at 800,000 BHCAs. The first softswitches offered no more than 250,000 BHCAs.

*Quality of Service (QoS)* The voice quality of the PSTN is the standard for telephone service. The Class 4 switch was the first deployment of TDM utilizing a 64 Kbps circuit (G.711 codec), which remains the standard to this day. Service on the PSTN is known for the absence of echo, crosstalk, latency, dropped or blocked calls, noise, or any other degradations of voice quality. Mainstream service providers in North America are reluctant to deploy equipment that offers voice quality at a lesser standard than the Class 4 or 5 switch.

One advantage service providers have had in delivering good QoS is that their legacy networks were designed specifically to deliver excellent voice quality and dependability in call setup and teardown. Historically, service providers have owned and operated their own proprietary networks over which they have had total control. Given end-to-end control, this has ensured good QoS for their subscribers. Good voice quality has long been a selling point for long-distance service providers.

Good QoS has been important in the legacy market. Service providers fear the loss of market share if they introduce a product in a converging market that does not deliver the same QoS as subscribers experienced in the legacy market. QoS is vitally important in legacy and converging markets.

---

[14]Nortel Networks. "Product Service Information-DMS300/250 System Advantage." www.nortel.com, 2001.

# Transport

The PSTN was built over the course of a century at a great expense. Developers have been obsessed over the years with getting the maximum number of conversations transported at the least cost in infrastructure possible. Imagine an early telephone circuit running from New York to Los Angeles. The copper wire, repeaters, and other mechanisms involved in transporting a conversation this distance were immense for its time. Hence, the early telephone engineers and scientists had to find ways to get the maximum number of conversations transported over this network. Through much research, different means were developed to wring the maximum efficiency from the copper wire infrastructure. Many of those discoveries translated into technologies that worked equally well when fiber optic cable came on the market. The primary form of transport in the PSTN has been TDM (described earlier in this chapter). In the 1990s, long-distance service providers (*interexchange carriers* or IXCs) and local service providers (*local exchange carriers* or LECs) have migrated those transport networks to *Asynchronous Transfer Mode* (ATM). ATM is the means for transport from switch to switch.

## Asynchronous Transfer Mode (ATM)

IXCs use high-speed switching systems to interconnect transmission lines. The key high-speed switching system used in IXC networks is ATM. ATM is a fast packet-switching technology that transports information through the use of small, fixed-length packets of data (53-byte cells).

The ATM system uses high-speed transmission facilities (155 Mbps/ OC-3 and above). OC-3 is the entry-level speed for commercial ATM. Higher speeds (such as OC-192) are used in backbone networks of IXCs and other specialized service providers. ATM service was developed to enable one communication technology (high-speed packet data) to provide for voice, data, and video service in a single offering. When an ATM circuit is established, a patch through multiple switches is set up and remains in place until the connection is completed.

The ATM switch rapidly transfers and routes packets to the predesignated destinations. To transfer packets to their destinations, each ATM switch maintains a database called a routing table. The routing table instructs the ATM switch as to which channel to transfer the incoming packet to and what priority should be given to the packet. The routing table is updated each time a connection is set up and disconnected. This enables the ATM switch to forward packets to the next ATM switch or destination point without spending much processing time.

The ATM switch also may prioritize or discard packets that it receives based on network availability (congestion). The ATM switch determines the prioritization and discard options by the type of channels and packets within the channels that are being switched by the ATM switch.

Three signal sources go through an ATM network to different destinations. The audio signal source (signal 1) is a 64 Kbps voice circuit. The data from the voice circuit is divided into short packets and sent to the ATM switch 1. ATM switch 1 looks in its routing table and determines the packet is destined for ATM switch 4, and ATM switch 4 adapts (slows down the transmission speed) and routes it to its destination voice circuit. The routing from ATM switch 1 to ATM switch 4 is accomplished by assigning the ATM packet a *virtual circuit identifier* (VCI) that an ATM switch can understand (the packet routing address). This VCI code remains for the duration of the communication.

The second signal source is a 384 Kbps Internet session. ATM switch 1 determines the destination of these packets is ATM switch 4 through ATM switch 3. The third signal source is a 1 Mbps digital video signal from a digital video camera. ATM switch 1 determines this signal is destined for ATM switch 4 for a digital television. In this case, the communication path is through ATM switches 1, 2, and 4.

## Optical Transmission Systems

At the physical layer, carriers use microwave or fiber optic cable to transport ATM packets containing voice and data from switch to switch. IXC backbone carrier facilities primarily use microwave and fiber transmission lines. Microwave systems offer a medium capacity of up to several hundred Mbps communications with a range of 20 to 30 miles between towers. Fiber optic communication systems offer a data transmission capacity of over one million Mbps (one million in a million bits per second).

Microwave transmission systems transfer signal energy through an unobstructed medium (no blocking buildings or hills) between two or more points. In 1951, microwave radio transmission systems became the backbone of the telecommunications infrastructure. Microwave systems require a transducer to convert signal energy of one form into electromagnetic energy for transmission. The transducer must also focus the energy (using an antenna dish) so it may launch the energy in the desired direction. Some of the electromagnetic energy that is transmitted by microwave systems is absorbed by the water particles in the air.

Although the extensive deployment of fiber optic cable has removed some of the need for microwave radio systems, microwave radio is still used in places that are hard to reach or are not cost effectively served by fiber cable, such as in developing countries. Fiber optic transmission is the transfer of information (usually in digital form) through the use of light pulses. Fiber optic transmission can be performed through glass fiber or through air. Fiber optic transmission lines are capable of extending up to 1,200 kilometers without amplifiers. Each fiber optic strand can carry up to 10 Gbps of optical channels and a fiber can have many optical channels (called DWDM). Each fiber cable can have many strands of fiber.

Fiber cable is relatively light and low cost, and it can be easily installed in a variety of ways. It does not experience distortion from electrical interference and this enables it to be installed on high-voltage power lines or in other places that have high levels of electromagnetic interference. Optical transmission systems use strands of glass or plastic fiber to transfer optical energy between points. For most optical transmission systems, the transmitting end-node uses a *light amplification through stimulated emission of radiation* (laser) device to convert digital information into pulsed light signals (amplitude modulation). The light signals travel down the fiber strand by bouncing (reflecting) off the sides of the fiber (called the cladding) until they reach the end of the fiber. The end of the fiber is connected to a photodetector that converts these light pulses back into their electrical signal form.

Synchronous optical transmission systems use a specific frame structure and the data transmission through the transmission line is synchronized to a precise clock. This eliminates the signaling overhead requirement for framing or timing alignment messages. The basic frame size used in optical transmission systems is 125 usec frames.

Optical transmission systems are characterized by their carrier level (OCX) where the basic carrier level 1 is 51.84 Mbps. Lower-level OC structures are combined to produce higher-speed communication lines. Different structures of OC are used in the world. The North American optical transmission standard is called *Synchronous Optical Network* (SONET) and the European (world standard) is *Synchronous Digital Hierarchy* (SDH).

Signals are applied to and are extracted from optical transmission systems using an *optical add/drop multiplexer* (OADM). The OADM is a network element that provides access to all or some subset *synchronous transport signal* (STS) line signals contained within an *optical carrier level N* (OC-N). The process used to direct a data signal or packet to a payload of an optical signal is called mapping. The mapping table is contained in the OADM. A copy of the OADM mapping is kept at other locations in the event of equipment failure. This allows the OADM to be quickly reprogrammed.

SDH is an international digital transmission format used in optical (fiber) standardized networks that is similar (but not identical) to SONET. SDH uses standardized synchronous transmissions according to ITU standards G.707, G.708, and G.709. These standards define data transfer rates, defined optical interfaces, and signal structure formats.

Some of the key differences between SONET and SDH include differences in overhead (control) bits and minimum transfer rates. The first level available in the SONET system is OC1 and is 51.84 Mbps. The first level in the SDH system starts at STM-1 and has a data transmission rate of 155.52 Mbps. SONET also multiplexes *synchronous transport signal level 1* (STS-1) to form multiple levels of STS. The SDH system divides the channels into multiple DS0s (64 Kbps channels). This is why the overhead signaling structures are different.

Table 2-6 shows the optical standards for both SONET and SDH. This table shows that the first common optical level between SONET and SDH

**Table 2-6**

*Optical transmission systems*

| OC level | Signal level | Sonet STS level | SDH STM level | DS0 (64 Kbps) channel |
|----------|--------------|-----------------|---------------|------------------------|
| OC-1 | 51.84 Mbps | STS-1 | | 672 |
| OC-2 | 103.68 Mbps | STS-2 | | 1,344 |
| OC-3 | 155.52 Mbps | STS-3 | STM-1 | 2,016 |
| OC-4 | 207.36 Mbps | STS-4 | STM-3 | 2,688 |
| OC-9 | 466.56 Mbps | STS-9 | STM-3 | 6,048 |
| OC-12 | 622.08 Mbps | STS-12 | STM-4 | 8,063 |
| OC-18 | 933.12 Mbps | STS-18 | STM-6 | 12,096 |
| OC-24 | 1.24416 Gbps | STS-24 | STM-8 | 16,128 |
| OC-36 | 1.86624 Gbps | STS-36 | STM-12 | 24,192 |
| OC-48 | 2.48832 Gbps | STS-48 | STM-16 | 32,256 |
| OC-96 | 4.976 Gbps | STS-96 | STM-32 | 64,512 |
| OC-192 | 9.953 Gbps | STS-192 | STM-64 | 129,024 |
| OC-256 | 13.219 Gbps | STS-256 | | |
| OC-768 | 52.877 Gbps | STS-786 | | |

is OC3 or STS-1. STS-x and STM-x are the standards that specify the electrical signal characteristics that are input to the respective optical encoding/multiplexing processes.

# Conclusion

This chapter has described the major components of the PSTN. By categorizing the diverse components of the PSTN into three simple elements, access, switching, and transport, a framework is provided for understanding how comparable elements of a softswitch solution replace those of the PSTN, enabling bottlenecks to be bypassed in the PSTN when delivering voice services to subscribers. Many concepts deployed in the PSTN have been translated into softswitched networks, including signaling, voice codecs, and transport. When this technology can be duplicated by startup technology providers and implemented by competitive service providers, competition to the local loop becomes possible.

# Softswitch Architecture or "It's the Architecture, Stupid!"

A softswitched network, like the *Public Switched Telephone Network* (PSTN), can be described in terms of its functions: access, switching, and transport. Where the two types of networks are decidedly different is in their architectures. The PSTN is a centralized architecture while a softswitch architecture offers a distributed architecture.

This chapter will describe the architecture for softswitch solutions, detailing the major components that comprise this new and flexible architecture. Although many of the rules of construction change drastically with the shift to a packet-switched client-server architecture, the basic functions remain the same. Much of the current architecture for softswitch networks reflects attempts to protect an investment in the legacy aspects of a network, and softswitch components bridge the legacy technology with next-generation technologies. As economic models dictate new strategies, legacy equipment will be replaced by softswitch architectures. New market entrants will have an advantage by deploying softswitch solutions from the start. Incumbents will be disrupted by an architecture that is cheaper, simpler, smaller, and more convenient to use.

# Softswitch and Distributed Architecture: A "Stupid" Network

Figure 3-1 illustrates the distributed architecture that is generally agreed upon as the model for softswitch networks. This model decouples the underlying packet-switching hardware from the call control, service logic, and new service creation. This distribution enables flexibility in hardware choices as well as the innovation of new services without requiring changes in the switching fabric or structure. This model also opens up the opportunities for third-party developers. The bottom layer is considered to be the bearer or transport plane, which physically transports both voice and data traffic. This plane consists of the media gateways in the softswitch solution.[1]

What makes this possible is the client-server architecture of softswitch, as opposed to the mainframe architecture of the Class 4 and 5 switches (see Figure 3-2). One advantage is that it enables a service provider to start

---

[1]Flynn, Christian. "Softswitches: The Brains Behind the Brawn." *Yankee Group*, May 2000. p. 3.

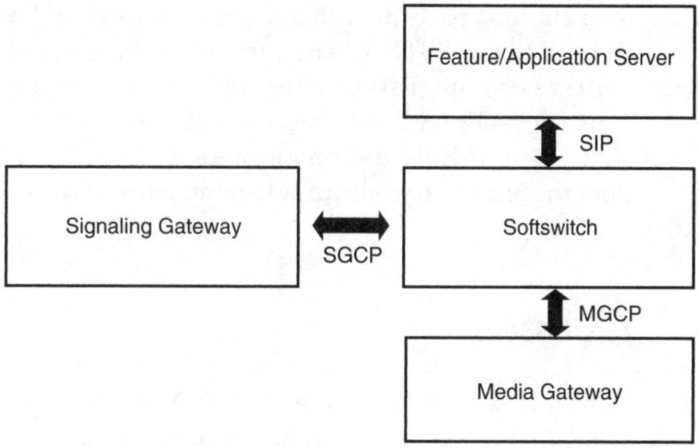

**Figure 3-1**
Softswitch
architecture
components

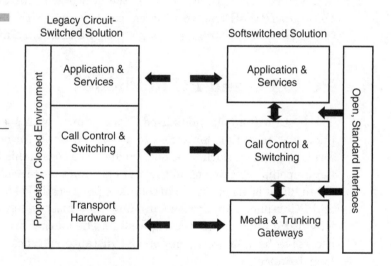

Feature/Application Server

SIP

Signaling Gateway

SGCP

Softswitch

MGCP

Media Gateway

**Figure 3-2**
Mainframe versus
softswitch client-
server architecture.
Note the fluidity of
the softswitch
architecture.

Legacy Circuit-
Switched Solution

Softswitched Solution

Proprietary, Closed Environment

Application &
Services

Call Control &
Switching

Transport
Hardware

Application &
Services

Call Control &
Switching

Media & Trunking
Gateways

Open, Standard Interfaces

small and grow with the demand, as opposed to a large upfront investment in a Class 4 switch.

In the fall of 1997, *Computer Telephony* printed a white paper entitled "Rise of the Stupid Network" authored by David Isenberg, a scientist at Bell Labs. In the paper, Isenberg points out that the Internet is the inverse of the PSTN in that the intelligence of the Internet resides at the periphery of the network, instead of residing at the core of the network as it does in the PSTN. Thus, Softswitch architecture reflects a "stupid" network. Softswitch is a sum of its parts distributed across an *Internet Protocol* (IP) network, as opposed to the PSTN where a few large, highly centralized Class 4 and 5 switches operate. This chapter outlines the components and ideology of a stupid network.[2] Softswitch can be considered a stupid solution as it utilizes distributed architecture (intelligence at the periphery), which is different than the "smart" or centralized architecture of the Class 4 and 5 switches.

# Access

Softswitch architecture, like the PSTN, can be described as having three elements: (1) access, that is, how a subscriber gains access to the network; (2) switching, how a call is controlled across the network; and (3) transport, or how a call is transported across the network. In the case of accessing a *Voice over IP* (VoIP) network, access can be gained either from an IP source (PC or IP phone) or from a legacy, analog handset via a media gateway.

## PC to PC and PC to Phone

The first VoIP applications used PCs equipped with speakers and microphones as terminals for access to a VoIP network. Initially, the *quality of service* (QoS) left much to be desired and, as a result, this form of access did not immediately catch on in the market. This service is often referred to as PC to PC. It is also possible to complete phone calls PC to Phone. PC-to-PC and PC-to-phone applications are now used most widely by consumers for long-distance bypasses (see Figure 3-3). The market driver for this form of access has been saving money on long distance, specifically on international long distance.

Although often touted as an enterprise telephony solution, the use of PC as a telephony terminal has not seized any significant market share. Even where the QoS was acceptable for the task, anthropological issues remained. PCs do not resemble telephones in appearance, feel, or function. This presents a psychological barrier to the user for using a PC as readily as a telephone handset.

---

[2]Isenberg, David. "Rise of the Stupid Network." *Computer Telephony,* August 1997. pg. 16–26. (Also see www.isen.com.)

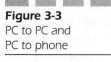

**Figure 3-3**
PC to PC and
PC to phone

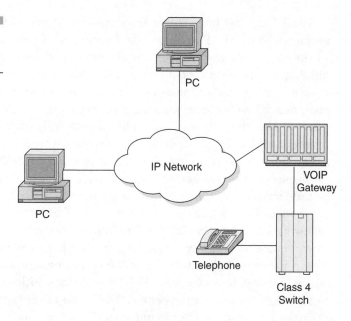

# IP Phones (IP Handsets)
# Phone-to-Phone VoIP

It didn't take the industry long to realize the benefits of a PC in a handset for use in VoIP, thus was born the IP phone. Early pioneers of this technology included e-tel and PingTel. The IP handset incorporates all the computer hardware necessary to make an IP phone call possible. Another strong advantage of the IP handset is that it removes anthropological objections to VoIP calls. The IP handset looks and functions like a telephone as opposed to a PC. IP handsets are stand-alone devices and present an IP-desktop-to-desktop solution.

IP handsets offer further benefits in that they do not require a gateway and its incumbent investment and management responsibilities. The chief advantage of an IP phone to an enterprise is that the phone requires a minimum of network configuration and management. Each employee equipped with an IP phone can take his or her phone anywhere on a network with no reconfiguration of the phone or the network. IP phone-equipped employees are potentially more productive because the *graphic user interface* (GUI) on the IP phone makes using features much easier than with a 12-button conventional telephone handset and its list of star codes.

The IP handset has its own IP address, which is recognized wherever it is connected on an IP network (see Figure 3-4). In an enterprise setting, a worker can disconnect his or her IP phone and move to another cubicle, building, or state, and the phone will function with no reprogramming necessary. In a legacy enterprise setting with a circuit-switched *private branch exchange* (PBX), tools for managing moves, adds, or changes tend to be difficult to use and, consequently, administrators learn only the basic management skills. This makes it very expensive to administer the switch. According to some estimates, it can cost as much as $300 to $500 per a PBX move, add, or change. For a Centrex line, it can take weeks for a change to be implemented by the telephone company.[3]

IP phones are available in two flavors (or VoIP protocols, to be covered in following chapters), H.323 and the *Session Initiation Protocol* (SIP), and they require no gateway. Physically, the IP phone connects to the network via an Ethernet connection (RJ-45). In a business environment, an Ethernet hub serves to concentrate VoIP phone lines, although in a legacy network there would be an expensive PBX. The advantage to a VoIP service provider is that it need not maintain a Class 4 or 5 switch or VoIP gateway.

**Figure 3-4**
IP phones on an IP
network

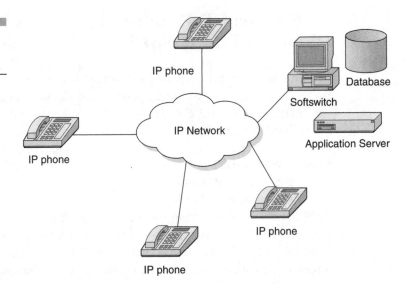

---

[3]"Next-Gen VoIP Services and Applications Using SIP and Java." A white paper from Applied Technologies Group, available at www.pingtel.com/docs/collateral_techguide_final.pdf.

In a legacy voice network, service providers must purchase and maintain large Class 4 and 5 switches. They must also measure their investment in a cost per DS0 (single phone line) on a switch that is further extrapolated into a cost per subscriber for the switch. In a VoIP network, it is generally assumed that the subscriber will purchase his or her own *customer premise equipment* (CPE). Hence, in such a next-generation network, the cost per DS0 for the service provider is $0.

SIP and Java programs also enable a whole new generation of applications that are impossible with circuit-switched telephony architectures. These applications can generally be divided into three categories: personal productivity applications, occupation-specific and industry-specific applications, and *web-telephony integration* (WTI) applications.

Most IP phones have *liquid crystal display* (LCD) screens with GUIs that enable expanded functions over a 12-button analog handset. With a conventional handset, the user must memorize long reams of number codes to perform functions such as conferencing, voice mail retrieval, call forwarding, and so on. For many users, this presents a psychological barrier that limits them to using only a handful of the features available on a PBX, thus preventing them from being as efficient in their communications as they could be. An IP phone with a GUI overcomes a number of these shortcomings by presenting the user with graphic choices to access their features.

A disadvantage to the IP phone is that, at the time of this writing, the IP phones on the market are very expensive relative to a conventional handset. IP phones from Cisco, Nortel, or PingTel cost at least $500, as opposed to a conventional PBX-connected handset at about $150 per handset. That high cost makes this technology unattractive to the residential market. However, an IP handset that was competitive in price to feature-rich analog or digital handsets would probably be very popular and would further the growth of IP telephony. Price competition will drive the price of IP phones to below $100 by late 2003.

In summary, the chief advantage of IP phones to a service provider is that they do not require the service provider to invest in a switch or gateway. In theory, the subscriber has covered that investment by buying the IP phone. Furthermore, service providers enjoy high margins on offering features, especially features not possible via circuit-switched telephones and networks. GUI interfaces on IP phones make these services easier to use, which can result in a greater marketability of those services for the service provider. In short, the voice network of the not-so-distant future will consist of IP phones that connect to IP networks where the intelligence for that service is provided by a softswitch located anywhere on the network.

# Media Gateways (a VoIP gateway switch)

The most successful commercial form of access has been the use of a VoIP gateway. The gateway provides a connection between an endpoint on a data network and the PSTN or switched-circuit network. The gateway translates between transmission formats and the communication procedures that are used on each side. Gateways can be provided as stand-alone devices or be integrated into other systems. In this form of access, an existing telephone handset interfaces a gateway either via a direct connection, a PBX, or a Class 5 switch. The gateway packetizes the voice and routes it over the IP network. Table 3-1 details the range of gateways and the scale of access they provide.

A media gateway can serve as both an access and limited-switching device. The media gateway resides on the edge of a network and simply interfaces between *Time Division Multiplexing* (TDM) and IP networks. It is here that analog or digital signals from a handset (PBX or Class 4 or 5 switch) are digitized (if analog), packetized, and compressed for transmission over an IP network. In the inverse, incoming calls are translated and decompressed from an IP network for reception on digital or analog telephone devices.[4] The media gateway interfaces directly with a TDM switch (PBX or Class 4 or 5).

The design of a gateway includes three key elements: an interface for the TDM side of the network (described in terms of DS0s or T1s), an interface for the packet side of the network (usually an Ethernet connection), and the necessary signal processing between these two sides. Signal pro-

**Table 3-1**

Gateways and their markets based on scalability

| Gateway | Scale |
| --- | --- |
| Residential/SOHO | 2 to 8 ports (DS0s) linking a telephone handset via an RJ-11 connector |
| Enterprise | From two ports to multiple T1/E1s connecting to a PBX or directly to a handset |
| Carrier-grade | High density, multiple T1/E1 line side, an OC-3, and a trunk side with 3,000-plus DS0s in a seven-foot rack |

[4]International Softswitch Consortium. "Enhanced Service Framework." Applications Working Group, www.Softswitch.org, 2001.

cessing is done with *digital signal processors* (DSP) on circuitboards designed to support voice. Signal processing functions include echo cancellation, coding/decoding of the analog signal with algorithms discussed in Chapter 2, "The Public Switched Telephone Network (PSTN)," such as G.711 or G.723.1, adapting the digitally encoded information into a series of IP datagrams, and transmitting those datagrams via a network to its ultimate destination.

On softswitch architecture, a media gateway can also be a part of the switching equation depending on the amount of intelligence contained in the gateway. The trend is toward less intelligence in the media gateway and more intelligence in the softswitch. In the early days of the VoIP industry, the gateway had to contain a good deal of intelligence to make calls possible. However, the evolution of gatekeeper technology into carrier-grade softswitch has drawn the intelligence out of the gateway and on to the softswitch.

Perhaps the most important issue for media gateways is their scalability. The density (the number of DS0s or ports in one chassis) determines its classification. Depending on its density, a media gateway falls into one of the three following classifications: residential or *small office home office* (SOHO), enterprise, or carrier grade.

**Residential or SOHO Gateways**　Figure 3-5 shows a residential grade of a VoIP gateway.

**Figure 3-5**
SOHO gateways in a
VOIP network

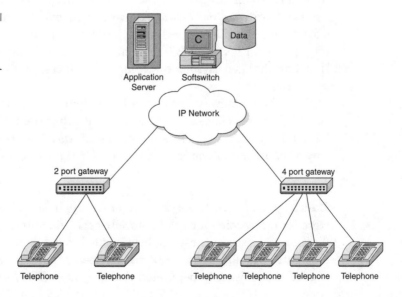

Application Server　　Softswitch　　Data

IP Network

2 port gateway　　　　4 port gateway

Telephone　Telephone　　Telephone　Telephone　Telephone　Telephone

**Figure 3-6**
Enterprise gateways
on a VoIP network

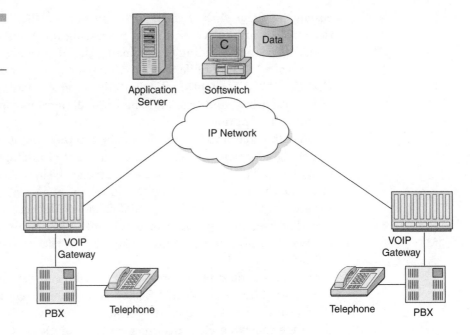

**Enterprise Gateways**  Enterprise gateways aggregate legacy telephone infrastructures for interface with VoIP networks (see Figure 3-6). This is usually done by connecting a gateway to the trunk side of a legacy PBX. Users retain their existing handsets. This has the effect of making VoIP indistinguishable to the end user. The business need not train its staff to use new hardware or software to work their existing telephone handsets. This option also offers investment protection in that the business retains its expensive PBX and PBX-associated telephone handsets. The only thing that has changed is that the company reduces or eliminates interoffice phone bills.

Enterprise gateways are usually configured in multiples of T1 or E1 cards in a single chassis that interface with the trunk side of the PBX. The T1 trunks connect to the line side of the gateway. The trunk side of the gateway is its Ethernet connection to a router if a router is not built into the gateway.

**Carrier-Grade Gateways**  The early applications for VoIP gateways were for international long-distance bypass and enterprise interoffice long distance. Success in these applications led to a demand for expanded density gateways for carrier operations (see Figure 3-7). These gateways needed to be densely populated (have enough DS0s or ports) enough to

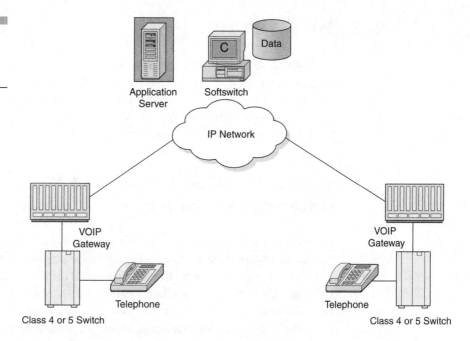

**Figure 3-7**
VoIP gateways in carrier-grade applications

interface with Class 4 and Class 5 switches (up to 100,000 DS0s in one node with an OC-3 trunk-side interface). These switches also had to offer the reliability to interface with a circuit switch that boasted five 9s of reliability.

Another requirement was that the switches be certified as being NEBS 3 compliant, a requirement for any platform to be installed in a central office. *Network Equipment Building Standards* (NEBS) addresses the physical reliability of a switch. It is contained in Telcordia specification SR 3580, an extensive set of rigid performance, quality, safety, and environmental requirements applicable to network equipment installed in a carrier's central office. Nearly all major carriers in North America require that equipment in their central offices or switching locations undergo rigorous NEBS testing. Tests include electrical safety, immunity from electromagnetic emissions, lightning and power faulting, and bonding and grounding evaluations. Equipment must meet physical standards as well, including temperature, humidity, altitude testing, fire resistance (usually by destructive burning), earthquake vibration resistance, and a battery of other rigid tests. NEBS compliance also means backup and disaster recovery strategies, including ensuring access to mirror sites, fire and waterproof storage facilities for critical databases and configuration backup information, and an *uninterruptible power supply* (UPS) to prevent network outages due to power failures.

# Switching

"Soft"switching is a break-through technology that has empowered VoIP to eclipse TDM as a telephony technology. Prior to the development of the softswitch, VoIP was handicapped by a lack of intelligence necessary to route calls across the network. Without this intelligence, an evolution to an IP alternative to the PSTN would not be possible.

## Softswitch (Gatekeeper and Media Gateway Controller)

A softswitch is the intelligence in a network that coordinates call control, signaling, and features that make a call across a network or multiple networks possible. Primarily, a softswitch performs call control. Call control performs call setups and teardowns. Once a call is set up, connection control ensures that the call stays up until it is released by the originating or terminating user. Call control and service logic refer to the functions that process a call and offer telephone features. Examples of call control and service logic functions include recognizing that a party has gone off hook and that a dial tone should be provided, interpreting the dialed digits to determine where the call is to be terminated, and determining if the called party is available or busy and finally, recognizing when the called party answers the phone and when either party subsequently hangs up, and recording these actions for billing.

A softswitch coordinates the routing of signaling messages between networks. Signaling coordinates actions associated with a connection to the entity at the other end of the connection. To set up a call, a common protocol must be used that defines the information in the messages and which is intelligible at each end of the network and across dissimilar networks. The main types of signaling a softswitch performs are peer to peer for call control and softswitch to gateway for media control. For signaling, the predominant protocols are SIP, *Signaling System 7* (SS7), and H.323. For media control, the predominant signaling protocol is the *Media Gateway Control Protocol* (MGCP).

As a point of introduction to softswitch, it is necessary to clarify the evolution to softswitch and define *media gateway controllers* (MGCs) and gatekeepers, which were the precursors to softswitch. MGCs and gatekeepers (essentially synonymous terms for the earliest forms of softswitch) were designed to manage low-density (relative to a carrier-grade solution) voice networks. MGC communicates with both the signaling gateway and the media gateway to provide the necessary call-processing functions. The MGC uses either MGCP or MEGACO/H.248 (described in a later chapter) for intergateway communications.

Gatekeeper technology evolved out of H.323 technology (a VoIP signaling protocol described in the next chapter). As H.323 was designed for local area networks where an H.323 gatekeeper would manage activities in a zone. A zone is a collection of one or more gateways managed by a single gatekeeper. A gatekeeper should be thought of as a logical function and not a physical entity. The functions of a gatekeeper are address translation (a name or email address for a terminal or gateway and a transport address) and admissions control (authorizing access to the network).

As VoIP networks got larger and more complex, management solutions with far greater intelligence became necessary. Greater call-processing power became necessary, as did the capability to interface signaling between IP networks with the PSTN (VoIP signaling protocols to SS7). Other drivers included the need to integrate features on the network and interface disparate VoIP protocols, thus was born the softswitch.

A significant market driver for softswitch is protocol intermediation, which is necessary to interface H.323 and SIP networks, for example. Another market driver for softswitch is to interface between the PSTN (SS7) and IP networks (SIP and H.323). Another softswitch function is the intermediation between media gateways of dissimilar vendors. Despite an emphasis on standards such as H.323, interoperability remains elusive. A softswitch application can overcome intermediation issues between media gateways. More information on VoIP protocols and signaling IP to PSTN is provided in following chapters.

Softswitch provides usage statistics to coordinate the billing, track operations, and administrative functions of the platform while interfacing with an application server to deliver value-added subscriber services. The softswitch controls the number and type of features provided. It interfaces with the application server to coordinate features (conferencing, call forwarding, and so on) for a call.

Physically, a softswitch is software hosted on a server chassis filled with IP boards and includes the call control applications and drivers.[5] Very simply, the more powerful the server, the more capable the softswitch. That server need not be colocated with other components of the softswitch architecture.

---

[5]Ibid.

# Signaling Gateway

Signaling gateways are used to terminate signaling links from PSTN networks or other signaling points. The SS7 signaling gateway serves as a protocol mediator (translator) between the PSTN and IP networks. That is, when a call originates in an IP network using H.323 as a VoIP protocol and must terminate in the PSTN, a translation from the H.323 signaling protocol to SS7 is necessary in order to complete the call. Physically, a signaling function can be embedded directly into the MGC or housed within a stand-alone gateway.

# Application Server

The application server accommodates the service and feature applications made available to a service provider's customers. Examples include call forwarding, conferencing, voice mail, forward on busy, and so on. Physically, an application server is a server loaded with a software suite that offers the application programs. The softswitch accesses, enables, and applies these application programs to the appropriate subscribers as needed (see Figure 3-2).

A softswitch solution emphasizes open standards as opposed to the Class 4 or 5 switch that has historically offered proprietary and closed environments. A carrier was a "Nortel shop" or a "Lucent shop." No components (hardware or software) from one vendor would be compatible with products from another vendor. Any application or feature on a DMS-250, for example, had to be a Nortel product or specifically approved by Nortel. This usually translates into less than competitive pricing for those components. Softswitch open standards are aimed at freeing service providers from vendor dependence and the long and expensive service development cycles of legacy switch manufacturers.

Concern as to whether a softswitch solution can transmit a robust feature list identical to that found on a Nortel DMS-250, Class 4 switch, for example, is an objection among service providers. Softswitch offers the advantage of enabling service providers to integrate third-party applications or even write their own while interoperating with the features of the PSTN via SS7. This is potentially the greatest advantage to a service provider presented by softswitch technology.

Features reside at the application layer in softswitch architecture. The interface between the call control layer and specific applications is the *application program interface* (API). Writing and interfacing an application with the rest of the softswitch architecture occurs in the service creation environment. This will be covered in greater detail in a following chapter.

## Applications for Softswitch

Where does a softswitch fit in a next-generation network? What does it do? How does it compare with PSTN switches? The following pages will note how softswitch technologies are deployed to replace TDM switches ranging from PBXs to Class 5 switches.

**IP PBX**   Perhaps the earliest application for VoIP was the installation of a VoIP gateway on the trunk side of a PBX. This gateway packetized the voice stream and routed it over an IP network (corporate WAN, UPN, Internet), which saved the business much money in long-distance transport costs. This solution used the existing PBX's set of features (conferencing, call forwarding, and so on). It also provided investment protection to the user by leveraging the legacy PBX into a VoIP solution. The intelligence in this solution was contained in software known as the gatekeeper. The gatekeeper was the precursor to the softswitch. That configuration is detailed in Figure 3-6.

Eventually, software developers devised a "soft" PBX, which could replace legacy PBXs (see Figure 3-8). These soft PBXs were considerably less expensive than a hardware PBX. They then came to be known as IP PBXs. An IP PBX can be thought of as an enterprise-grade softswitch.

**Figure 3-8**
IP PBX, also known as soft PBX

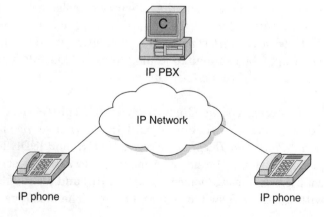

IP PBX

IP Network

IP phone                IP phone

**IP Centrex**   Just as the Centrex model followed the PBX in circuit switching, it does the same in packet switching. Shortly after IP PBXs began to catch on in the market, the *Regional Bell Operating Companies* (RBOCs) began to realize a threat to their circuit-switched Centrex services from VoIP applications. Centrex accounted for about 15 percent of all business lines, and many subscribers were locked into five-year contracts with the RBOCs. As these contracts expired in the late 1990s, many customers were actively evaluating less expensive alternatives to Centrex. If large companies could route their interoffice voice traffic over a corporate *wide area network* (WAN) using an IP PBX, what would be the demand for their circuit-switched Centrex services? With this threat in mind, IP Centrex services arrived on the market. Centrex is a set of specialized business solutions (primarily, but not exclusively, for voice service) where the equipment providing the call control and service logic functions is owned and operated by the service provider and hence is located on the service provider's premises. Since Centrex frees the customer from the costs and responsibilities of major equipment ownership, Centrex can be thought of as an outsourcing solution.

In traditional Centrex service (analog Centrex and ISDN Centrex), call control and service logic reside in a Class 5 switch located in the central office. The Class 5 switch is also responsible for transporting and switching the electrical signals that carry the caller's speech or other information (such as faxes).

IP Centrex refers to IP telephony solutions where Centrex service is offered to a customer who transmits his voice calls to the network as packetized streams across an IP network. One benefit is increased utilization of access capacity. In IP Centrex, a single broadband access facility is used to carry the packetized voice streams for many simultaneous calls. In analog Centrex, one pair of copper wires is needed to serve each analog telephone station, regardless of whether the phone has an active call. Once the phone is not engaged in a call, the bandwidth capacity of those wires is unused. An ISDN *basic rate interface* (BRI) can support two simultaneous calls (that is, 128 Kbps), but similar to analog lines, an idle BRI's bandwidth capacity cannot be used to increase the corporate LAN's interconnection speed.

**IP Centrex Using Class 5 Switch Architecture**   In this platform, existing Class 5 switches support IP Centrex service in addition to traditional *Plain Old Telephone Service* (POTS) and ISDN lines. This is accomplished through the use of a media gateway (as described earlier in this chapter) on the *customer premises* (CPE) and a GR-303 gateway colocated with the Class 5 switch (see Figure 3-9). The media gateway can be of any

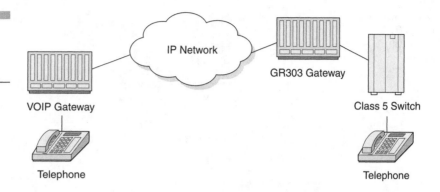

**Figure 3-9**
IP Centrex using a
Class 5 switch and
GR-303 interface

size from an IP phone to a carrier-grade media gateway. The media gateway connects to the switch as if it were a digital loop carrier system. (Digital loop carriers use protocols such as GR-303 to deliver POTS and ISDN signaling information to switches for longer-than-average loops.) The GR-303 gateway translates any signaling information it receives from the customer's media gateway and depacketizes the voice stream for delivery to the switch. Similarly, it translates signaling messages from the switch into the IP telephony protocol (H.323, SIP, or MGCP) and packetizes the voice stream for transmission to the customer's media gateway.

The customer's media gateway performs comparable functions for the standard telephone sets that it supports. As a result, the GR-303 gateway, the customer's media gateway, and the IP network connecting them appear to the Class 5 switch as an ordinary *digital loop carrier* (DLC) system, and the telephone sets connected to the customer gateway appear to the switch as ordinary phone lines. Because the IP Centrex solution is treated as a DLC system by the Class 5 switch, the switch is able to deliver the same features to IP Centrex users that it delivers to analog and ISDN Centrex users. Consequently, an extensive set of features is immediately available to IP Centrex users without needing to upgrade the Class 5 switch.

**IP Centrex Using Softswitch Architecture**   In a different approach to IP Centrex, the Class 5 switch is replaced by a softswitch (see Figure 3-10). A softswitch is a telephony application running on a large, high-availability server on the network. Like the Class 5 switch, the softswitch provides call control and service logic. Unlike the Class 5 switch, the softswitch is not involved in the transport or switching of the packetized voice stream. The softswitch and the IP Centrex CPE (customer media

**Figure 3-10**
IP Centrex with
softswitch

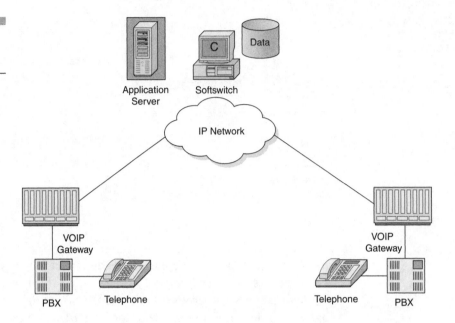

gateways and IP phones) signal one another over a packet network using an IP telephony protocol, such as H.323 or SIP. After it receives call setup information, the softswitch determines where the called party resides. If the called party is a member of the Centrex group, then the softswitch instructs the originating media gateway (or IP phone) and the terminating media gateway (or IP phone) to route the packetized voice streams directly to one another. Consequently, the voice stream never leaves the corporate LAN/WAN. If the called party is served by the PSTN, then the softswitch instructs the originating media gateway (or IP phone) to route the packetized voice stream to a trunking gateway.

The trunking gateway has traditional interoffice facilities to Class 4 or Class 5 switches in the PSTN. It packetizes/depacketizes the voice stream so that it can be transmitted over these circuit-switched facilities. The trunking gateway works in conjunction with a signaling gateway, which is used to exchange SS7 messages with the PSTN. Both the trunking and signaling gateways receive their instructions from the softswitch.[6]

## Class 4 Replacement Softswitch

The next step in scale for the VoIP industry and tangentially the softswitch industry was Class 4 replacement. The origins of Class 4 replacement softswitch solutions lay in the long-distance bypass industry. Long-distance bypass operators used VoIP gateways for international transport. This technology enabled them to be competitive relative to the "big three" long-distance companies. Part of that success was due to the fact that they were able to avoid paying into international settlements (described later in this book). Initially, these service providers used enterprise-grade media gateways that interfaced with TDM switches in the PSTN. Technical challenges for these operators arose as their business flourished and demand grew. First, the media gateways were not dense enough for the levels of traffic they were handling. Second, the gateways that controlled these gateways were also limited in their capability to handle ever-increasing levels of traffic over these networks. Thirdly, international traffic called for interfacing different national variants of SS7 signaling (each nation has its own variant).

In short, market demand dictated that a more scalable and intelligent solution be offered in the long-distance bypass industry. That solution came in the form of what is known as a Class 4 replacement softswitch solution, comprised of more densely populated gateways managed with greater intelligence than an MGC. The first applications involved installing a dense gateway on the trunk side of a Class 4 switch, such as a Nortel DMS-250. As in the PBX scenario, the media gateway packetized the voice stream coming out of the Class 4 switch and routed it over an IP network, saving the service provider money on long-distance transport.

The next step in the evolution of Class 4 replacement softswitch was the removal of the circuit-switched Class 4 switch from that architecture. That is, the Class 5 switch connected directly to a media gateway that routed the call over an IP network. The call control, signaling, and features were controlled by a softswitch, and the Class 4 switch was replaced in its entirety.

For the purposes of this book, it is assumed that the arena of competition is similar to a scenario where Class 4 switches (DSM-250s from Nortel) are connected to an IP backbone and long-distance traffic is transported via that IP backbone.[7] At this service provider, softswitch, as a Class 4 replacement switch, competes directly with the Class 4 switch. Figures 3-11 and 3-12 illustrate this evolution in architecture.

---

[6]IP-Centrex.org web site (www.ip-centrex.org).

[7]Interview with John Hill, Network Sales Engineer, Williams Communication, October 31, 2001.

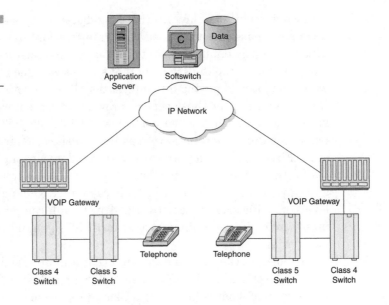

**Figure 3-11**
Architecture of Class
4 switches with VoIP
gateways

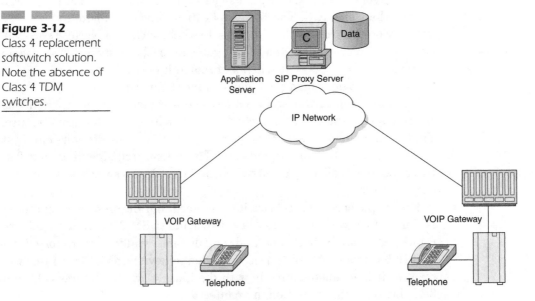

**Figure 3-12**
Class 4 replacement
softswitch solution.
Note the absence of
Class 4 TDM
switches.

## Class 5 Replacement Softswitch

The next level of progression in the development of softswitch technologies
was the Class 5 replacement (see Figure 3-13). This has led the most exciting debate over softswitch. The ability of the softswitch industry to replace

**Figure 3-13**
Class 5 replacement
softswitch solution

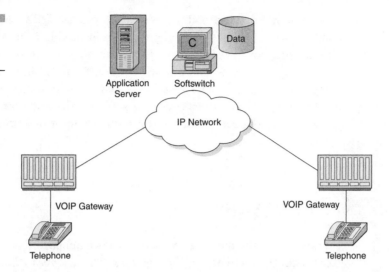

the Class 5 marks the final disruption of the legacy telecommunication infrastructure. A Class 5 switch can cost tens of millions of dollars and require at least half a city block in real estate. The evolution of a successful Class 5 replacement softswitch has staggering implications for the world's local telephone service providers.

From the early days of the telephone industry, it was assumed that the cost of deploying local phone service with its copper pair access and local phone switches (most recently, a Class 5) was so expensive that only a monopoly could affect this economy of scale and scope. Enter a Class 5 replacement softswitch that does not cost tens of millions of dollars nor require a centrally located and expensive central office, and the barriers to entry and exit crumble. The result is that new market entrants may be able to effectively compete with quasi-monopolistic incumber service providers. This is potentially disruptive to incumbent local service providers and their Class 5 switch vendors.

Objections to a Class 5 replacement softswitch solution include the need for E911 and Communication Assistance for law Enforcement Act of 1994 (CALEA). This will be addressed in a later chapter. Another objection is the perception that softswitch cannot match Class 5 features. A 5ESS Class 5 switch from Lucent Technologies is reported to have some 3,500 features that have been developed over a 25-year time frame. This features debate will be addressed in a later chapter. At the time of this writing, a number of successful Class 5 replacement softswitch installations have taken place and this segment of the industry is growing rapidly.

In summary, the softswitches that replace PBXs and Class 4 and 5 switches (including Centrex) are differentiated in their scale. That is, their processing power is measured by the number of busy-hour call attempts or calls per second they can handle. Other differentiating factors include their capability to handle features from a feature server and interface disparate signaling protocols. A softswitch is software that rides on a server. The limitations are the complexity of the software and the processing power of the server.

# Transport

Transport is the means by which voice is transported across the network. It connects the switches in the network. The *Memorandum for Final Judgement* (MFJ) in 1984 broke up AT&T and opened the long-distance market to competition. In essence, this opened the transport market. Simply put, three modes of voice transport are in use today: IP, *Asynchronous Transfer Mode* (ATM), and TDM.

## Legacy, Converging, and Converged Architecture

VoIP and softswitch technologies rose out of the economic necessity for long-distance service providers to switch to the least expensive means of transport, which is IP. By bypassing TDM and ATM networks, long-distance service providers greatly reduced their costs of long-distance transport, which made them more competitive and more profitable than their TDM- or ATM-equipped competitors. Softswitch and VoIP got their beginnings in the transport aspect of the network. Long-distance service providers needed an intelligence that would perform call control over the IP network they used for their transport. In addition, softswitch needed to interface SS7 to the IP network, and finally it had to control the transmission of features across the IP network. Thus was born the Class 4 replacement softswitch. The following paragraphs detail its evolution from the legacy network to the converging network and use in the converged network.

Service providers speak of a telecommunications market where voice, data, and perhaps video and other broadband services are provided over a single network, presumably based on IP. The subscriber consequently enjoys highly efficient IP services desktop to desktop. This is called a *converged* network. The vast majority of the Class 4 and 5 switch market was designed and installed when voice and data were handled via separate channels. These are referred to as *legacy* networks (see Figure 3-14). This book refers to the transition network as a *converging* network (see Figure 3-15). To define the markets for Class 4 and 5 versus softswitch, it is important to understand that legacy markets apply to legacy networks where voice and data are separate networks. A converging market applies to converging networks, where, in most instances, the legacy infrastructure of Class 4 and 5 switches remains at the periphery of the network while the core of the network is IP, which provides efficient voice transport. A converged market applies to a converged network where voice and data are handled on one network (see Figure 3-16).

In the converged market, voice switching is performed by "classless" switches. This is because the limitations of geography defined a Class 5 switch as providing local service and a Class 4 switch as providing long distance. If geography is irrelevant, then a Class designation is irrelevant.

## IP Networks

This book will not address the many benefits of IP over other means of transport. Why is IP becoming the preferred means of transport in the data world and soon to be the preferred means of transport for voice? Because it is available almost everywhere. Originally designed for

**Figure 3-14**
Legacy networks: separate voice and data networks

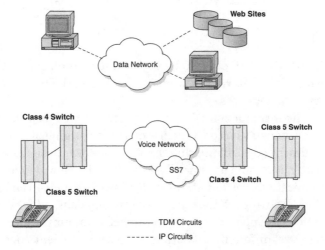

**Figure 3-15**
Converging
networks: a mix of
circuit- and packet-
switched networks
and technologies

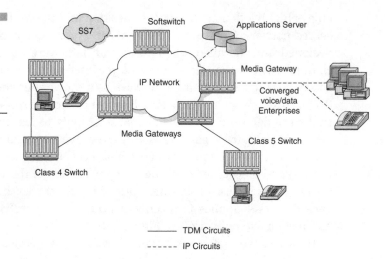

**Figure 3-16**
A converged network
is one network for
voice and data.

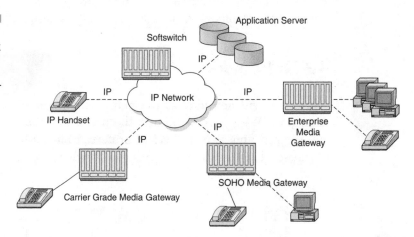

DARPANET, IP grew to be the dominant protocol for the "network of networks," the Internet. It went on to serve as the transport medium of choice for data networks.

The introduction of VoIP in the late 1990s focused the telecommunications industry on convergence, or the converging of voice and data networks as described earlier in this chapter. TDM was designed strictly for voice and does not handle data nearly as well as IP. New demands in business communications for email, instant messaging, video conferencing, and World Wide Web applications dictate a growing demand for data-based communications as opposed to circuit-switched communications. Wiring and managing two different networks are also inefficient, not to mention costly.

Converging voice and data networks has the following advantages: lower equipment costs, the integration of voice and data applications, lower bandwidth requirements, and the widespread availability of IP.

Large service providers (RBOCS and IXCs) in today's market operate the PSTN, the Frame Relay network, the ATM network, the IP network, the ISDN and DSL overlay networks, and, for some service providers, wireless and X.25 networks. Each of these networks requires separate maintenance, installation, operations, accounting, configuration, and provisioning systems with the necessary personnel and network management centers to monitor them. In contrast, IP offers the promise of replacing all those disparate systems with one central, efficient network.

Today virtually every enterprise is already using the *Transmission Control Protocol* (TCP)/IP protocol stack. IP is ubiquitous throughout the business world. Every personal computer supports IP. IP is used for LANs and WANs, dial-up Internet access, and handheld computers, and it is the preferred technology for wireless web and other next-generation applications. The result is a major opportunity to link systems and networks together through a combination of public and private IP infrastructures in much the same way as telephone systems combine PBXs, private voice networks, and the PSTN. With the possible exception of the SS7 signaling protocols, IP is the world's most widely deployed protocol. It also offers the world's only universal addressing scheme.

The integration of voice and data services is only now beginning to come into fruition. Service providers now offer an expansion of high-margin services that differentiate them from circuit-switched service providers. Examples include follow-me services, "click-n-call," and the use of an email address as the equivalent of a telephone number. Video conferencing over IP is another benefit of converged voice and data networks. Economic models emphasizing these advantages will be covered in a later chapter.

The anticipation of almost unlimited demand for IP-based services drove the formation of IP backbone providers in the late 1990s. Unfortunately, the anticipated level of demand did not materialize as hoped, resulting in oversupply of IP backbone capacity. This has proven to be a boon for IP consumers, including VoIP service providers.

## ATM

At the time of this writing, ATM is the predominant means of long-distance voice transport in North America. It replaced TDM based on its greater efficiency over TDM and has yet to be superceded by IP, given its reputation for good QoS, which is the product of the capabilities built into its adaptation layer.

The ATM adaptation layer at the ingress switch examines customer traffic as it arrives at the switch. Then, based on the type of traffic, the ATM adaptation layer classifies the traffic per its QoS requirements. These are based on three parameters: whether the traffic is connectionless or connection-oriented, whether it requires a fixed or variable bit rate, and whether or not an explicit timing relationship exists between the sending and receiving devices. Following that process, the ATM ingress switch then assigns a series class to the cells that make up the traffic stream and transmits them into the network, knowing that ATM's highly reliable connection-oriented transport architecture and each switch's capability to interpret and respond to the assigned service class will ensure that the QoS mandate of the sending device will be accommodated on a network-wide basis. Many incumbent service providers now find themselves faced with devising a means of interworking their ATM and IP networks as they try to shift more traffic to the much more efficient IP networks.

## TDM

TDM was described in Chapter 2 in conjunction with the PSTN. It is synonymous with circuit-switched, legacy telephone networks. The term softswitch comes up most often when used in context with VoIP. It should be noted that comparable architecture is possible with a TDM network. Instead of the mainframe computer architecture of the Class 4 and 5, a TDM softswitch solution uses the client-server architecture of softswitch. Instead of VoIP gateways, it uses TDM gateways, which can be just as scalable as a Class 4 or 5 switch.

# Conclusion

The PSTN is comprised of three main elements: access, switching, and transport. The implementation of the 1984 MFJ has opened the long-distance markets for competition in the United States. In essence, the MFJ opened transport in the PSTN to competition. The Telecommunications Act of 1996 was designed to open the elements of switching and access for competition, but it failed miserably.

Softswitch technologies enable the telecommunications market to open the switching aspect to competition. Softswitch also enables a service provider to bypass the central office and Class 5 and Class 4 switches, which, given their extreme expense, technical complexity, and closed standards have protected monopolistic telcos for decades.

Part of the reason so many of the *Competitive Local Exchange Carriers* (CLECs) of the telecom boom of the late 1990s failed is that they were forced to compete with incumbent carriers by deploying the same expensive switches as the incumbents. The incumbents had the advantage of installed switches that were being depreciated often on 40-year schedules. A later chapter will describe the economic advantages of softswitch over Class 4 or Class 5, but suffice it to say that softswitch is cheaper both in purchase price and operating expenses than the circuit switches.

Under the Telecommunications Act of 1996, when CLECs attempted to contract with the incumbents to use incumbent switching facilities under provisions of the Telecommunications Act of 1996, they were met with legal stonewalling tactics by the incumbents. This was followed by outright sabotage by incumbent employees of CLEC equipment colocated in incumbent central offices. The only path to competition in telecommunications markets is for each service provider to own and operate their own switching facilities. Softswitch, given its lower costs of ownership and operation coupled with its distributed architecture, makes it possible for new market entrants to compete favorably with incumbent service providers.

Softswitch is disruptive to legacy networks (RBOCS and IXCs) and equipment vendors (makers of PBXs and Class 4 and 5 switches) in that elements of softswitch are cheaper, simpler, smaller, and more convenient to use than their predecessors in the network. This chapter has outlined the potentially disruptive technologies relative to the functions of the network elements that they replace. In a later chapter, this book will address the specifics of how these technologies are disruptive and who is disrupted by their introduction into the network and industry.

This leaves the final aspect of the PSTN, access, to be opened for competition. Access, like switching, was to be opened to competition by the Telecommunications Act of 1996. Competitors found the same roadblocks to access thrown up by the incumbents as they did to switching. The ownership of twisted-pair copper by quasi-monopolistic incumbent telcos leaves the vast majority of the network closed to realistic competition.

This book will address "last mile" solutions in a later chapter. Once the "last mile" issue has been solved, consumers will enjoy true competition in the local loop. Some solutions include wireless applications, Ethernet, free space lasers, and power lines. The benefits of competition in the local loop will not be limited to lower prices on POTS. The benefits of competition in or to the local loop will be lower prices for POTS, expanded voice services (follow me, conferencing, and so on), and expanded data and video services (broadband Internet, video conferencing from home, and video on demand). All these technologies are available now. For a variety of economic, business, political and technical reasons, quasi-monopolistic RBOCs have failed to deliver these services to their subscribers. The deployment of these services to the last mile will have a strong positive impact on local economic development while expanding people's quality of life through improved education, entertainment, and communication.

# Voice over Internet Protocol

# What Is VoIP?

Softswitch is a product driven by the need to incorporate intelligence into *Voice over Internet Protocol* (VoIP) networks, interface IP networks, and the *Public Switched Telephone Network* (PSTN) and to coordinate features across networks. As outlined in the previous chapter, the first applications of softswitch were the gatekeepers (aka gateway controllers) that were incorporated in networks of VoIP gateways. In order to better understand a softswitched network, it is necessary to dissect VoIP down to the protocol level. Many volumes on VoIP can be found on the book market, and this book will not attempt to cover it in detail. The importance of VoIP protocols relative to softswitch is that they are the building blocks that make VoIP possible.

# Origins

In November 1988, Republic Telcom (yes, one "e") of Boulder, Colorado, received patent number 4,782,485 for a "Multiplexed Digital Packet Telephone System." The plaque from the Patent and Trademark Office describes it as follows: "A method for communicating speech signals from a first location to a second location over a digital communication medium comprising the steps of: providing a speech signal of predetermined bandwidth in analog signal format at said first location; periodically sampling said speech signal at a predetermined sampling rate to provide a succession of analog signal samples; representing said analog signal samples in a digital format thereby providing a succession of binary digital samples; dividing said succession of binary digital samples into groups of binary digital samples arranged in a temporal sequence; transforming at least two of said groups of binary digital samples into corresponding frames of digital compression."

Republic and its acquiring company, Netrix Corporation, applied this voice over data technology to the data technologies of the times (X.25 and Frame Relay) until 1998 when Netrix and other competitors introduced VoIP onto their existing voice over data gateways. Although attempts at internet telephony had been done from a software-only perspective, commercial applications were limited to using voice over data gateways that could interface the PSTN to data networks. Voice over data applications were popular in enterprise networks with offices spread across the globe (eliminated international interoffice long-distance bills), offices where no PSTN existed (installations for mining and oil companies), and for long-distance bypasses (legitimate and illegitimate).

The popularity and applications of VoIP continued to grow. VoIP accounted for 6 percent of all international long-distance traffic in 2001.[1] Six percent may not seem like an exciting sum, but given a mere 3 years from the introduction of a technology to capturing 6 percent of a trillion dollar, 100-year-old industry, it is clear that VoIP will continue to capture more market share.

# How Does VoIP Work?

Softswitch is increasingly considered to be almost synonymous with VoIP. However, it also works with *Time Division Multiplexing* (TDM) and *Asynchronous Transfer Mode* (ATM) networks. The first process in an IP voice system is the digitization of the speaker's voice. The next step (and the first step when the user is on a handset connected to a gateway using a digital PSTN connection) is typically the suppression of unwanted signals and compression of the voice signal. This has two stages. First, the system examines the recently digitized information to determine if it contains voice signal or only ambient noise and discards any packets that do not contain speech. Secondly, complex algorithms are employed to reduce the amount of information that must be sent to the other party. Sophisticated codecs enable noise suppression and the compression of voice streams. Compression algorithms include G.723, G.728, and G.729.

Following compression, voice must be packetized and VoIP protocols added. Some storage of data occurs during the process of collecting voice data, since the transmitter must wait for a certain amount of voice data to be collected before it is combined to form a packet and transmitted via the network. Protocols are added to the packet to facilitate its transmission across the network. For example, each packet will need to contain the address of its destination, a sequencing number in case the packets do not arrive in the proper order, and additional data for error checking. Because IP is a protocol designed to interconnect networks of varying kinds, substantially more processing is required than in smaller networks. The network addressing system can often be very complex, requiring a process of encapsulating one packet inside another and, and as data moves along, repackaging, readdressing, and reassembling the data.

---

[1]"TeleGeography 2002—Global Traffic Statistics and Commentary," *TeleGeography*, 2001, www.TeleGeography.com.

When each packet arrives at the destination computer, its sequencing is checked to place the packets in the proper order. A decompression algorithm is used to restore the data to its original form, and clock synchronization and delay-handling techniques are used to ensure proper spacing. Because data packets are transported via the network by a variety of routes, they do not arrive at their destination in order. To correct this, incoming packets are stored for a time in a jitter buffer to wait for late-arriving packets. The length of time in which data are held in the jitter buffer varies depending on the characteristics of the network.

In IP networks, a percentage of the packets can be lost or delayed, especially in periods of congestion. Also, some packets are discarded due to errors that occurred during transmission. Lost, delayed, and damaged packets result in a substantial deterioration of voice quality. In conventional error-correction techniques used in other protocols, incoming blocks of data containing errors are discarded, and the receiving computer requests the retransmission of the packet; thus, the message that is finally delivered to the user is exactly the same as the message that originated. As VoIP systems are time-sensitive and cannot wait for retransmission, more sophisticated error detection and correction systems are used to create sound to fill in the gaps. This process stores a portion of the incoming speaker's voice and then, using a complex algorithm to approximate the contents of the missing packets, new sound information is created to enhance the communication. Thus, the sound heard by the receiver is not exactly the sound transmitted, but rather portions of it have been created by the system to enhance the delivered sound.[2]

The previous description details the movement of voice over the IP medium. The rest of the chapter will describe the building blocks of a VoIP network: its protocols.

# Protocols Related to VoIP

The softswitch revolution was made possible by the emergence of voice over data, more specifically, VoIP. It should be noted here that softswitch solutions use TDM and ATM. However, the consensus in the industry is that the future is an IP network ultimately dictating a VoIP solution. Before outlining softswitch solutions, it will first be necessary to understand VoIP. VoIP

---

[2]Report to Congress on Universal Service, CC Docket No. 96-45, White Paper on IP Voice Services, March 18, 1998 (www.von.org/docs/whitepap.pdf).

is best understood as a collection of the protocols that make up its mechanics. Those protocols are loosely analogous to the PSTN that is broken down into three categories: access, switching, and transport. Simply put, three categories of protocols are relevant to VoIP: signaling, routing, and transport.

Signaling (roughly analogous to the switching function described in the last two chapters) protocols (H.323 and *Session Initiation Protocol* [SIP]) set up the route for the media stream or conversation. Gateway control protocols such as the *Media Gateway Control Protocol* (MGCP) and MEGACO (also signaling protocols) establish control and status in media and signaling gateways.

Routing (using the *User Datagram Protocol* [UDP] and *Transmission Control Protocol* [TCP]) and transporting (*Real-Time Transport Protocol* [RTP]) the media stream (conversation) once the route of the media stream has been established are the functions of routing and transport protocols. Routing protocols such as UDP and TCP could be compared to the "switching" function described in Chapter 2, "The Public Switched Telephone Network (PSTN)," and Chapter 3, "Softswitch Architecture or 'It's the Architecture, Stupid!'"

RTP would be analogous to the "transport" function outlined in earlier chapters describing the PSTN and softswitch architectures. The signaling and routing functions establish what route the media stream will take. The routing protocols deliver the bits, that is, the conversation.

## Signaling Protocols

The process of setting up a VoIP call is roughly similar to that of a circuit-switched call made on the PSTN. A media gateway must be loaded with the parameters to allow proper media encoding and the use of telephony features. Inside the media gateway is an intelligent entity known as an endpoint. When the calling and called parties agree on how to communicate and the signaling criteria is established, the media stream over which the packetized voice conversation will flow is established. Signaling establishes the virtual circuit over the network for that media stream. Signaling is independent of the media flow. It determines the type of media to be used in a call. Signaling is concurrent throughout the call. Two types of signaling are currently popular in VoIP: H.323 and SIP.[3]

---

[3]Douskalis, Bill. *IP Telephony The Integration of Robust VoIP Services*. New York: Prentice Hall, 2000.

**Figure 4-1**
Signaling and
transport protocols
used in VoIP

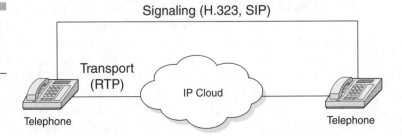

Figure 4-1 details the relationship between signaling and media flow. This relationship between transport and signaling is very similar to the PSTN in that *Signaling System 7* (SS7) is out-of-channel signaling, such as that used in VoIP.

**H.323**    H.323 is the *International Telecommunication Union—Telecommunications Standardization Sector* (ITU-T) recommendation for packet-based multimedia communication. H.323 was developed before the emergence of VoIP. As it was not specifically designed for VoIP, it has faced a good deal of competition from a competing protocol, SIP, which was designed specifically for VoIP. However, it has enjoyed a first-mover advantage and a considerably installed base of H.323 VoIP networks now exists.

H.323 is comprised of a number of subprotocols. It uses protocol H.225.0 for registration, admission, status, call signaling, and control. It also uses protocol H.245 for media description and control, terminal capability exchange, and general control of the logical channel carrying the media stream(s). Other protocols make up the complete H.323 specification, which presents a protocol stack for H.323 signaling and media transport. H.323 also defines a set of call control, channel setup and codec specifications for transmitting real-time video and voice over networks that don't offer guaranteed service or *quality of service* (QoS). As a transport, H.323 uses RTP, an *Internet Engineering Task Force* (IETF) standard designed to handle the requirements of streaming real-time audio and video via the Internet.[4] H.323 was the first VoIP protocol for interoperability among the early VoIP gateway/gatekeeper vendors. Unfortunately, the promise of interoperability between diverse vendors platforms did not materialize with the adoption of H.323. Given the gravity of this protocol, it will be covered in a separate following chapter.

---

[4]Ibid., pg. 9.

The H.323 standard is a cornerstone technology for the transmission of real-time audio, video, and data communications over packet-based networks. It specifies the components, protocols, and procedures providing multimedia communication over packet-based networks. Packet-based networks include IP-based (including the Internet) or *Internet packet exchange* (IPX)-based *local area networks* (LANs), *enterprise networks* (ENs), *metropolitan area networks* (MANs), and *wide area networks* (WANs). H.323 can be applied in a variety of mechanisms: audio only (IP telephony); audio and video (videotelephony); audio and data; and audio, video, and data. H.323 can also be applied to multipoint-multimedia communications. H.323 provides myriad services and therefore can be applied in a wide variety of areas: consumer, business, and entertainment applications.

**Interworking with Other Multimedia Networks**   The H.323 standard specifies four kinds of components, which, when networked together, provide the point-to-point and point-to-multipoint multimedia communication services: terminals, gateways, gatekeepers, and *multipoint control units* (MCUs).

**Terminals**   Used for real-time bidirectional multimedia communications, an H.323 terminal can either be a *personal computer* (PC) or a stand-alone device running an H.323 and the multimedia applications. It supports audio communications and can optionally support video or data communications. Because the basic service provided by an H.323 terminal is audio communications, an H.323 terminal plays a key role in IP-telephony services. The primary goal of H.323 is to interwork with other multimedia terminals. H.323 terminals are compatible with H.324 terminals on SCN and wireless networks, H.310 terminals on *Broadband Integrated Services Digital Network* (B-ISDN), H.320 terminals on ISDN, H.321 terminals on B-ISDN, and H.322 terminals on guaranteed QoS LANs. H.323 terminals may be used in multipoint conferences.

**Gateways**   A gateway connects two dissimilar networks. An H.323 gateway provides connectivity between an H.323 network and a non-H.323 network. For example, a gateway can connect and provide communication between an H.323 terminal and TDM networks This connectivity of dissimilar networks is achieved by translating protocols for call setup and release, converting media formats between different networks, and transferring information between the networks connected by the gateway. A gateway is not required, however, for communication between two terminals on an H.323 network.

**Gatekeepers**   A gatekeeper can be considered the brain of the H.323 network. It is the focal point for all calls within the H.323 network. Although they are not required, gatekeepers provide important services such as addressing, authorization, and authentication of terminals and gateways, bandwidth management, accounting, billing, and charging. Gatekeepers may also provide call-routing services.

**Multipoint Control Units (MCUs)**   MCUs provide support for conferences of three or more H.323 terminals. All terminals participating in the conference establish a connection with the MCU. The MCU manages conference resources, negotiates between terminals for the purpose of determining the audio or video *coder/decoder* (codec) to use, and may handle the media stream. The gatekeepers, gateways, and MCUs are logically separate components of the H.323 standard but can be implemented as a single physical device.

**H.323 Zone**   An H.323 zone is a collection of all terminals, gateways, and MCUs managed by a single gatekeeper. A zone includes at least one terminal and may include gateways or MCUs. A zone has only one gatekeeper. A zone may be independent of network topology and may be comprised of multiple network segments that are connected using routers or other devices.

Additional protocols specified by H.323 are listed in the following sections. H.323 is independent of the packet network and the transport protocols over which it runs and does not specify them. They are audio codecs; video codecs; H.225 *registration, admission, and status* (RAS); H.225 call signaling; H.245 control signaling; RTP; and the *Real-Time Control Protocol* (RTCP).

**Audio Codec**   An audio codec encodes the audio signal from the microphone for transmission on the transmitting H.323 terminal and decodes the received audio code that is sent to the speaker on the receiving H.323 terminal. Because audio is the minimum service provided by the H.323 standard, all H.323 terminals must have at least one audio codec support, as specified in the ITU-T G.711 recommendation (audio coding at 64 Kbps). Additional audio codec recommendations such as G.722 (64, 56, and 48 Kbps), G.723.1 (5.3 and 6.3 Kbps), G.728 (16 Kbps), and G.729 (8 Kbps) may also be supported.

**Video Codec**   A video codec encodes video from the camera for transmission on the transmitting H.323 terminal and decodes the received video code that is sent to the video display on the receiving H.323 terminal. Because H.323 specifies the support of video as optional, the support of video codecs is optional as well. However, any H.323 terminal providing video communications must support video encoding and decoding as specified in the ITU-T H.261 recommendation.

**H.225 Registration, Admission, and Status (RAS)**  RAS is the protocol between endpoints (terminals and gateways) and gatekeepers. RAS is used to perform registration, admission control, bandwidth changes, and status, and to disengage procedures between endpoints and gatekeepers. An RAS channel is used to exchange RAS messages. This signaling channel is opened between an endpoint and a gatekeeper prior to the establishment of any other channels.

**H.225 Call Signaling**  The H.225 call signaling is used to establish a connection between two H.323 endpoints. This is achieved by exchanging H.225 protocol messages on the call-signaling channel, which is opened between two H.323 endpoints or between an endpoint and the gatekeeper.

**H.245 Control Signaling**  H.245 control signaling is used to exchange end-to-end control messages governing the operation of the H.323 endpoint. These control messages carry information related to the following: capabilities exchange, the opening and closing of logical channels used to carry media streams, flow-control messages, general commands, and indications.

**Real-Time Transport Protocol (RTP) and H.323**  RTP, a transport protocol, provides end-to-end delivery services of real-time audio and video. Whereas H.323 is used to transport data over IP-based networks, RTP is typically used to transport data via the UDP. RTP, together with UDP, provides transport-protocol functionality. RTP provides payload-type identification, sequence numbering, timestamping, and delivery monitoring. UDP provides multiplexing and checksum services. RTP can also be used with other transport protocols.

**Real-Time Transport Control Protocol (RTCP) and H.323**  RTCP is the counterpart of RTP that provides control services. The primary function of RTCP is to provide feedback on the quality of the data distribution. Other RTCP functions include carrying a transport-level identifier for an RTP source, called a canonical name, which is used by receivers to synchronize audio and video.

**SIP**   SIP is a text-based signaling protocol used for creating and controlling multimedia sessions with two or more participants. It is a client-server protocol transported over TCP. SIP can interwork with gateways that provide signaling protocols and media translations across dissimilar network segments such as PSTN to IP networks. SIP uses text-based messages, much like HTTP. SIP addressing is built around either a telephone number or a web host name. In the case of a web host name, the SIP address is based on a *uniform resource locator* (URL). The URL is translated into an IP address through a *domain name server* (DNS). SIP also negotiates the features and capabilities of the session at the time the session is established.[5] SIP plays such a pivotal role in the evolution of VoIP and softswitching that it must be covered in the separate, following chapter.

**Gateway Control Protocols**   The most immediate attraction to VoIP is to save money on long-distance transport. To date, it has been impractical to route VoIP "desktop to desktop," meaning interworking between PSTN and IP networks must be facilitated. This is done with a gateway. The two most applied gateways are the media gateway and the signaling gateway. Media gateways interconnect dissimilar networks. In this case, they connect the PSTN to IP networks. To do this successfully, they must intermediate both signaling and transport between the two dissimilar networks (PSTN and IP). Media gateways coordinate call control and status. Gateway control protocols are signaling protocols.

**MGCP**   MGCP is the protocol used to intermediate the *Media Gateway Controller* (MGC, also known as a call agent) and the media gateway. MGCP was developed by the IETF and details the commands and parameters that are passed between the MGC and the telephony gateway to be controlled.

MGCP assumes a call control architecture where the call control intelligence is outside the gateways and is handled by external call control elements. The MGCP assumes that these call control elements, or call agents, will synchronize with each other to send coherent commands to the gateways under their control. MGCP is a master/slave protocol, where the gateways are expected to execute commands sent by the call agents.

The purpose of MGCP is to send commands from the call agent to a media gateway (see Chapter 3 for descriptions of media gateways). MGCP defines both endpoints and connections. Endpoints are sources or sinks of

---

[5]Ibid.

data and can be either physical (such as an interface terminating a digital trunk or analog line) or virtual (such as a designated audio source). An example of a virtual endpoint is an audio source in an audio-content server. The creation of physical endpoints requires hardware installation, while the creation of virtual endpoints can be done by software. Endpoint identifiers have two components, the domain name of the gateway that is managing the endpoint, and a local name within that gateway. Examples of physical endpoints include interfaces on gateways that terminate a trunk connected to a PSTN switch (Class 5 or Class 4) or an analog *Plain Old Telephone Service* (POTS) connection to a phone, key system, PBX, and so on. MGCP sends commands from the call agent to a media gateway. MGCP defines both endpoints and connections.

Connections can be either point to point or multipoint in nature. Further, connections are grouped into calls, where one or more connections can belong to one call. A point-to-point connection is an association between two endpoints with the purpose of transmitting data between these endpoints. Once this association is established for both endpoints, a data transfer between these endpoints can take place. A multipoint connection is established by connecting the endpoint to a multipoint session. For point-to-point connections, the endpoints of a connection could be in separate gateways or in the same gateway.

The connections and calls are established by the actions of one or more call agents. The information communicated between call agents and endpoints is either events or signals. An example of an event would be a telephone going off hook, while a signal may be the application of a dial tone to an endpoint. These events and signals are grouped into what are called packages, which are supported by a particular type of endpoint. One package may support events and signals for an analog line, while another package may support a group of events and signals for video lines.

As long as media gateways are interfacing with analog or PSTN connections to IP networks, MGCP will be the controlling protocol. MGCP will continue to be an integral element in any softswitch architecture.[6] Figure 4-2 details the function of MGCP in softswitch architecture and Table 4-1 outlines the signaling protocols and their softswitch components.

---

[6]Internet Engineering Task Force. *Request For Comments* (RFC) 2705, Media Gateway Control Protocol, October 1999.

**Figure 4-2**
The relationship between signaling protocols and softswitch architecture components

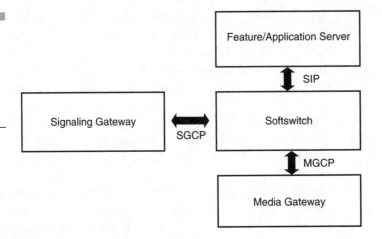

**Table 4-1**

Signaling protocols and associated softswitch components

| Network elements | Signaling protocol elements |
| --- | --- |
| Media gateway—softswitch | H.323 |
| Media gateway—media gateway | BICC |
| Media gateway—softswitch | MGCP |
| Softswitch—SS7 network | ISUP, TCAP |
| Softswitch or application server | SIP |

Source: Internet Telephony

**BICC**   *Bearer Independent Call Control* (BICC) is a newer protocol used for media-gateway-to-media-gateway communications. It provides a means of supporting narrowband ISDN services across a broadband backbone network without impacting the interfaces to the existing N-ISDN network and end-to-end services. The BICC call control signaling protocol is based on N-ISUP signaling.[7]

[7]Ratta, Greg. "Bearer Independent Call Control and its application of H.248 in public networks." ITU-T IP/MediaComm Workshop 2004. www.itu.int/itudoc/itu-t/com13/ipexpert/ipmedia/71396.html. April 26, 2001, pg. 3.

**SS7-Related Protocols**   In order for IP telephony networks to interoperate with the PSTN, they must interface with SS7. Softswitch solutions must include *ISDN User Part* (ISUP) and *Transaction Capabilities Application Part* (TCAP). ISUP, defined by ITU-T Q.761 and Q.764, is the call control part of the SS7b protocol. ISUP is an SS7 protocol for signaling the parameters and procedures to set up and tear down circuit-switched voice calls between a softswitch/signaling gateway and an STP. ISUP determines the procedures for call setup and teardown on the SS7 network.[8]

TCAP is a peer protocol to ISUP in the SS7 protocol hierarchy for end-to-end signaling not associated with call setup or specific trunks in the PSTN network. Some of its main uses are toll-free 800 number translations for routing across the network and *local number portability* (LNP). It also provides the interface between databases and SCPs.[9] TCAP provides services to any number of application parts. Common application parts include the *Intelligent Network Application Part* (INAP) and the *Mobile Application Part* (MAP).[10] This book will address the interworkings of SS7 and VoIP networks in greater detail in a following chapter.

# Routing Protocols

VoIP is routed over an IP network via routers. In order to deliver the best QoS, voice packets must be given priority over data packets. That means communicating to routers which packets have what priority. Router operations involve several processes. First, the router creates a routing table to gather information from other routers about the optimum path for each packet. This table may be static in that it is constructed by the router according to the current topology and conditions. Dynamic routing is considered a better technique as it adapts to changing network conditions. The router uses a metric of the shortest distance between two endpoints to help determine the optimum path. The router determines the least cost (most efficient) path from the origin to the destination.

Two algorithms are used to determine the least cost route. They are distance vector and link state. Protocols that make use of these algorithms are called *Interior Gateway Protocols* (IGPs). The *Routing Information Protocol*

---

[8]Newton, Harry. *Newton's Telecom Dictionary, 16th edition*. New York: CMP Books. pg. 486.

[9]Douskalis, pg. 9.

[10]Collins, Michael. *Carrier Grade Voice Over IP*. New York: McGraw-Hill, 2001. pg. 311.

(RIP) is an IGP based on the distance vector algorithm and the *Open Shortest Path First* (OSPF) protocol is an IGP based on the link state algorithm. Where one network needs to communicate with another, it uses an *Exterior Gateway Protocol* (EGP). One example of an EGP is the *Border Gateway Protocol* (BGP).

**Routing Information Protocol (RIP)**   RIP is a distance vector protocol that uses a hop count (the number of routers it passes through on its route to its destination) as its metric. RIP is widely used for routing traffic on the Internet and is an IGP, which means that it performs routing within a single autonomous system. EGPs, such as BGP, perform routing between different autonomous systems. RIP itself evolved as an Internet routing protocol, and other protocol suites use modified versions of RIP.

**Open Shortest Path First (OSPF) Protocol**   OSPF is a routing protocol developed for IP networks by the IGP Working Group of the IETF. The Working Group was formed in 1988 to design an IGP based on the *Shortest Path First* (SPF) algorithm for use on the Internet. Similar to the *Interior Gateway Routing Protocol* (IGRP), OSPF was created because in the mid-1980s RIP was increasingly incapable of serving large, heterogeneous internetworks.

OSPF has two primary characteristics. The first is that the protocol is open, which means that its specification is in the public domain. The OSPF specification is published as the IETF's RFC 1247. The second principal characteristic is that OSPF is based on the SPF algorithm, which sometimes is referred to as the Dijkstra algorithm, named for the person credited with its creation.

OSPF is a link-state routing protocol that calls for the sending of *link-state advertisements* (LSAs) to all other routers within the same hierarchical area. Information on attached interfaces, the metrics used, and other variables are included in OSPF LSAs. As OSPF routers accumulate link-state information, they use the SPF algorithm to calculate the shortest path to each node.

**SPF Algorithm**   The SPF routing algorithm is the basis for OSPF operations. When an SPF router is powered up, it initializes its routing-protocol data structures and then waits for indications from lower-layer protocols that its interfaces are functional. After a router is assured that its interfaces are functioning, it uses the OSPF Hello protocol to acquire neighbors, which are routers with interfaces to a common network. The router sends hello packets to its neighbors and receives their hello packets. In addition to helping acquire neighbors, Hello packets also act as keep-alives to let routers know that other routers are still functional.

Each router periodically sends an LSA to provide information on a router's adjacencies or to inform others when a router's state changes. By comparing established adjacencies to link states, failed routers can be detected quickly and the network's topology altered appropriately. From the topological database generated from LSAs, each router calculates a shortest-path tree, with itself as root. The shortest-path tree, in turn, yields a routing table.[11]

**Border Gateway Protocol**  BGP performs interdomain routing in TCP/IP networks. BGP is an EGP, which means that it performs routing between multiple autonomous systems or domains and exchanges routing and reachability information with other BGP systems. BGP was developed to replace its predecessor, the now obsolete EGP, as the standard exterior gateway-routing protocol used on the global Internet. BGP solves serious problems with EGP and scales to Internet growth more efficiently.

BGP performs three types of routing: interautonomous system routing, intra-autonomous system routing, and passthrough autonomous system routing. Interautonomous system routing occurs between two or more BGP routers in different autonomous systems. Peer routers in these systems use BGP to maintain a consistent view of the internetwork topology. BGP neighbors communicating between autonomous systems must reside on the same physical network. The Internet serves as an example of an entity that uses this type of routing because it is comprised of autonomous systems or administrative domains. Many of these domains represent the various institutions, corporations, and entities that make up the Internet. BGP is frequently used to provide path determination to provide optimal routing within the Internet.

Intra-autonomous system routing occurs between two or more BGP routers located within the same autonomous system. Peer routers within the same autonomous system use BGP to maintain a consistent view of the system topology. BGP also is used to determine which router will serve as the connection point for specific external autonomous systems. Once again, the Internet provides an example of interautonomous system routing. An organization, such as a university, could make use of BGP to provide opti-mal routing within its own administrative domain or autonomous system. The BGP protocol can provide both inter- and intra-autonomous system routing services.

---

[11]Cisco Systems. "Open Shortest Path First." A white paper from Cisco Systems, www.cisco.com.

Passthrough autonomous system routing occurs between two or more BGP peer routers that exchange traffic across an autonomous system that does not run BGP. In a passthrough autonomous system environment, the BGP traffic does not originate within the autonomous system in question and is not destined for a node in the autonomous system. BGP must interact with whatever intra-autonomous system routing protocol is being used to successfully transport BGP traffic through that autonomous system.

**BGP Routing**   As with any routing protocol, BGP maintains routing tables, transmits routing updates, and bases routing decisions on routing metrics. The primary function of a BGP system is to exchange network-reachability information, including information about the list of autonomous system paths, with other BGP systems. This information can be used to construct a graph of autonomous system connectivity from which routing loops can be pruned and with which autonomous-system-level policy decisions can be enforced.

Each BGP router maintains a routing table that lists all feasible paths to a particular network. The router does not refresh the routing table, however. Instead, routing information received from peer routers is retained until an incremental update is received. BGP devices exchange routing information upon an initial data exchange and after incremental updates. When a router first connects to the network, BGP routers exchange their entire BGP routing tables. Similarly, when the routing table changes, routers send the portion of their routing table that has changed. BGP routers do not send regularly scheduled routing updates, and BGP routing updates advertise only the optimal path to a network.

BGP uses a single routing metric to determine the best path to a given network. This metric consists of an arbitrary unit number that specifies the degree of preference of a particular link. The BGP metric typically is assigned to each link by the network administrator. The value assigned to a link can be based on any number of criteria, including the number of autonomous systems through which the path passes, stability, speed, delay, or cost.[12]

**Resource Reservation Protocol (RSVP)**   The *Resource Reservation Protocol* (RSVP) is a network control protocol that enables Internet applications to obtain special *qualities of service* (QoSs) for their data flows.

---

[12]Cisco Systems. "Border Gateway Protocol." A white paper from Cisco Systems. www.cisco.com, June 1999.

RSVP is not a routing protocol; instead, it works in conjunction with routing protocols and installs the equivalent of dynamic access lists along the routes that routing protocols calculate. RSVP occupies the place of a transport protocol in the OSI Model seven-layer protocol stack. The IETF is now working toward standardization through an RSVP working group. RSVP operational topics discussed in this chapter include data flows, QoS, session startup, reservation style, and soft state implementation.

In RSVP, a data flow is a sequence of messages that have the same source, destination (one or more), and QoS. QoS requirements are communicated through a network via a flow specification, which is a data structure used by internetwork hosts to request special services from the internetwork. A flow specification often guarantees how the internetwork will handle some of its host traffic.

RSVP supports three traffic types: best-effort, rate-sensitive, and delay-sensitive. The type of data-flow service used to support these traffic types depends on the QoS implemented. The following paragraphs address these traffic types and associated services. For information regarding QoS, refer to the appropriate section later in this chapter.

Best-effort traffic is traditional IP traffic. Applications include file transfers, such as mail transmissions, disk mounts, interactive logins, and transaction traffic. The service supporting best-effort traffic is called best-effort service. Rate-sensitive traffic is willing to give up timeliness for guaranteed rate. Rate-sensitive traffic, for example, might request 100 Kbps of bandwidth. If it actually sends 200 Kbps for an extended period, a router can delay traffic. Rate-sensitive traffic is not intended to run over a circuit-switched network; however, it usually is associated with an application that has been ported from a circuit-switched network (such as ISDN) and is running on a datagram network (IP).

An example of such an application is H.323 videoconferencing, which is designed to run on ISDN (H.320) or ATM (H.310) but is found on the Internet. H.323 encoding is a constant rate or nearly constant rate, and it requires a constant transport rate. The RSVP service supporting rate-sensitive traffic is called *guaranteed bit-rate service*. Delay-sensitive traffic is traffic that requires the timeliness of delivery and varies its rate accordingly. MPEG-II video, for example, averages about 3 to 7 Mbps depending on the amount of change in the picture. As an example, 3 Mbps might be a picture of a painted wall, although 7 Mbps would be required for a picture of waves on the ocean. MPEG-II video sources send key and delta frames. Typically, 1 or 2 key frames per second describe the whole picture, and 13 or 28 frames describe the change from the key frame. Delta frames are usually substantially smaller than key frames. As a result, rates vary quite a bit from frame to frame. A single frame, however, requires delivery within a frame time or the codec is unable to do its job. A specific priority must be negotiated for delta-frame traffic. RSVP services supporting delay-sensitive traffic are referred to as controlled-delay service (nonreal time service) and predictive service (real-time service).

In the context of RSVP, QoS is an attribute specified in flow specifications that is used to determine the way in which data interchanges are handled by participating entities (routers, receivers, and senders). RSVP is used to specify the QoS by both hosts and routers. Hosts use RSVP to request a QoS level from the network on behalf of an application data stream. Routers use RSVP to deliver QoS requests to other routers along the path(s) of the data stream. In doing so, RSVP maintains the router and host state to provide the requested service.[13]

## Transport Protocols

Figure 4-1 shows that the actual VoIP conversation takes place over a channel separate from the signaling channel. The following paragraphs describe RTP, the transport protocol.

**Real-Time Protocol (RTP)**    RTP is the most popular of the VoIP transport protocols. It is specified in RFC 1889 under the title of "RTP: A Transport Protocol for Real Time Applications." This RFC describes both RTP and another protocol, RTCP. As the name would suggest, these two protocols are necessary to support real-time applications like voice and video. RTP operates on the layer above UDP, which does not avoid packet loss or guarantee the correct order for the delivery of packets. RTP packets overcome those shortcomings by including sequence numbers that help applications using RTP to detect lost packets and ensure packet delivery in the correct order.

RTP packets include a timestamp that gives the time when the packet is sampled from its source media stream. This timestamp assists the destination application to determine the synchronized playout to the destination user and to calculate delay and jitter, two very important detractors of voice quality. RTP does not have the capacity to correct delay and jitter, but it does provide additional information to a higher-layer application so that the application can make determinations as to how a packet of voice or data is best handled.

---

[13]Cisco Systems. "Resource Reservation Protocol." A white paper from Cisco Systems, www.ciscosystems.com, June 1999.

RTCP provides a number of messages that are exchanged between session users and that provide feedback regarding the quality of the session. The type of information includes details such as the numbers of lost RTP packets, delays, and interarrival jitter. As voice packets are transported in RTP packets, RTCP packets transfer quality feedback. Whenever an RTP session opens, an RTCP session is also opened. That is, when a UDP port number is assigned to an RTP session for the transfer of media packets, another port number is assigned for RTCP messages.

**RTP Payloads**   RTP carries the digitally encoded voice by taking one or more digitally encoded voice samples and attaching an RTP header to provide RTP packets, which are made up of an RTP header and a payload of the voice samples. These RTP packets are sent to UDP, where a UDP header is attached. This combination then goes to IP where an IP header is attached and the resulting IP datagram is routed to the destination. At the destination, the headers are used to pass the packet up the stack to the appropriate application.

**RTP Headers**   RTP carries the carried voice in a packet. The RTP payload is comprised of digitally coded samples. The RTP header is attached to this payload and the packet is sent to the UDP layer. The RTP header contains the necessary information for the destination to reconstruct the original voice sample.

**RTP Control Protocol (RTCP)**   RTCP enables exchanges of control information between session participants with the goal of providing quality-related feedback. This feedback is used to detect and correct distribution issues. The combination of RTCP and IP multicast enables a network operator to monitor session quality. RTCP provides information on the quality of an RTP session. RTCP empowers network operators to obtain information about delay, jitter, and packet loss and to take corrective action where possible to improve quality.

# IPv6

The previous discussion assumed the use of *Internet Protocol version 4* (IPv4), the predominant version of IP in use today. A new version, IPv6, is now coming on the market. The explosion of Internet addresses necessitates the deployment of IPv6. IPv6 makes possible infinitely more addresses than IPv4. Enhancements offered by IPv6 over IPv4 include

- **Expanded address space**  Each address is allocated 128 bits instead of 32 bits in IPv4.
- **Simplified header format**  This enables easier processing of IP datagrams.
- **Improved support for headers and extensions**  This enables greater flexibility for the introduction of new options.
- **Flow-labeling capability**  This enables the identification of traffic flows for real-time applications.
- **Authentication and privacy**  Support for authentication, data integrity, and data confidentiality are supported at the IP level rather than through separate protocols or mechanisms above IP.

# Conclusion

This chapter addressed the building blocks of VoIP. It will be necessary in future chapters to understand many of the concepts contained in this chapter. Just as the PSTN and softswitch networks can be broken down into the three elements of access, switching, and transport, VoIP can be summarized as a study of three types of protocols: signaling, routing, and transport. The proper selection of VoIP signaling protocols for a network is an issue of almost religious proportions among network builders. Although protocols will continue to evolve and new protocols will emerge, those addressed in this chapter will constitute the predominant structure of softswitch architecture for the next few years.

# SIP: Alternative Softswitch Architecture?

If the worldwide *Public Switched Telephone Network* (PSTN) could be replaced overnight, the best candidate architecture, at the time of this writing, would be based on *Voice over IP* (VoIP) and the *Session Initiation Protocol* (SIP). Much of the VoIP industry has been based on offering solutions that leverage existing circuit-switched infrastructure (such as VoIP gateways that interface a *private branch exchange* [PBX] and an *Internet Protocol* [IP] network). At best, these solutions offer a compromise between circuit- and packet-switching architectures with resulting liabilities of limited features, expensive-to-maintain circuit-switched gear, and questionable *quality of service* (QoS) as a call is routed between networks based on those technologies. SIP is an architecture that potentially offers more features than a circuit-switched network.

SIP is a signaling protocol. It uses a text-based syntax similar to the *Hypertext Transfer Protocol* (HTTP) like that used in web addresses. Programs that are designed for the parsing of HTTP can be adapted easily for use with SIP. SIP addresses, known as SIP *uniform resource locators* (URLs) take the form of web addresses. A web address can be the equivalent of a telephone number in an SIP network. In addition, PSTN phone numbers can be incorporated into an SIP address for interfacing with the PSTN. An email address is portable. Using the proxy concept, one can check his or her email from any Internet-connected terminal in the world. Telephone numbers, simply put, are not portable. They only ring at one physical location. SIP offers a mobility function that can follow subscribers to whatever phone they are nearest to at a given time.

Like H.323, SIP handles the setup, modification, and teardown of multimedia sessions, including voice. Although it works with most transport protocols, its optimal transport protocol is the *Real Time Protocol* (RTP) (refer to Chapter 3, "Softswitch Architecture or 'It's the Architecture, Stupid!'" for more information on RTP). Figure 5-1 shows how SIP functions as a signaling protocol while RTP is the transport protocol for a voice conversation. SIP was designed as a part of the *Internet Engineering Task Force* (IETF) multimedia data and control architecture. It is designed to interwork with other IETF protocols such as the *Session Description Protocol* (SDP), RTP, and the *Session Announcement Protocol* (SAP). It is described in the IETF's RFC 2543. Many in the VoIP and softswitch industry believe that SIP will replace H.323 as the standard signaling protocol for VoIP.

SIP is part of the IETF standards process and is modeled upon other Internet protocols such as the *Simple Mail Transfer Protocol* (SMTP) and HTTP. It is used to establish, change, and tear down (end) calls between one or more users in an IP-based network. In order to provide telephony services, a number of different standards and protocols must come together—

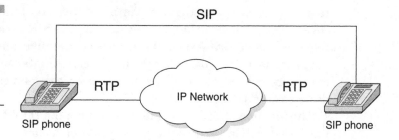

**Figure 5-1**
SIP is a signaling
protocol and RTP
transports the
conversation
protocol.

specifically to ensure transport (RTP), provide signaling with the PSTN, guarantee voice quality (*Resource Reservation Setup Protocol* [RSVP]), provide directories (*Lightweight Directory Access Protocol* [LDAP]), authenticate users (*Remote Access Dial-In User Service* [RADIUS]), and scale to meet anticipated growth curves.

# What Is SIP?

SIP is focused on two classes of network entities: clients (also called *user agents* [UAs]) and servers. VoIP calls on SIP to originate at a client and terminate at a server. Types of clients in the technology currently available for SIP telephony include a PC loaded with a telephony agent or an SIP telephone. Clients can also reside on the same platform as a server. For example, a PC on a corporate *wide area network* (WAN) might be the server for the SIP telephony application, but it may also be used as a user's telephone (client).

## SIP Architecture

SIP is a client-server architecture. The client in this architecture is the UA, which interacts with the user. It usually has an interface towards the user in the form of a PC or an IP phone (an SIP phone in this case). Four types of SIP servers exist. The type of SIP server used determines the architecture of the network. Those servers are the *user agent server* (UAS), the redirect server, the proxy server, and a registrar.

**SIP Calls via a UA Server**   A UA server accepts SIP requests and contacts the user. A response from the user to the UA server results in an SIP response representing the user. An SIP device will function as both a *UA client* (UAC) and as a UA server. As a UAC, the SIP device can initiate SIP requests. As a UA server, the device can receive and respond to SIP requests. As a standalone device, the UA can initiate and receive calls that empowers SIP to be used for peer-to-peer communications. Figure 5-2 describes the UA server.

The function of SIP is best understood via the HTTP model upon which it is based. SIP is a request/response protocol. A client is an SIP entity that generates a request. A server is an SIP entity that receives requests and returns responses. When a web site is desired, the client generates a request by typing in the URL, such as www.mcgrawhill.com. The server upon which the web site is hosted responds with McGraw-Hill's web page. SIP uses the same procedure. The UA sending the request is known as a UAC, and the UA returning the response is the UAS. This exchange is known as an SIP transaction.

Per Figure 5-2, a call initiates with the UA of the calling party sending an INVITE command caller@righthere.org to the called party, callee@theotherside.com. The UA for the caller has translated the name for the called party into an IP address via a *Domain Name System* (DNS) query accessible via their own domain. The INVITE command is sent to an SIP *User Datagram Protocol* (UDP) port and contains information such as media format and the From, To, and Via information. The TRYING informational response (180) from the calling party's call agent is analogous to the Q.931 CALL PROCEEDING message used in the PSTN, indicating the call is being routed. In the direct call model, a TRYING response is unlikely, but for proxy and redirect models it is used to monitor call progress. SIP

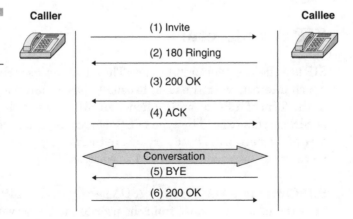

**Figure 5-2**
SIP UA server to UA server call

**Table 5-1**

*SIP signaling methods*

| SIP method | Description |
|---|---|
| INVITE | The first message sent by a calling party to invite users to participate in a session. It contains information in the SIP header that identifies the calling party, call ID, called party, and call and sequence numbers. It indicates a call is being initiated. When a multiple choice of SDP parameters is offered, the ones chosen are returned with the success (200) code in the response message. |
| ACK | This is used to acknowledge the reception of a final response to an INVITE. A client originating an INVITE request issues an ACK request when it receives a final response for the INVITE, providing a three-way handshake. |
| OPTIONS | This queries a server about its capabilities, including which methods and which SDPs it supports. This determines which media types a remote user supports before placing the call. |
| BYE | This is used to abandon sessions. In two-party sessions, abandonment by one of the parties implies that the session is terminated. A return BYE from the other party is not necessary. |
| CANCEL | This method cancels pending transactions. The CANCEL method identifies the call via the call ID, call sequence, and To and From values in the SIP header. |
| REGISTER | These requests are sent by users to inform a server about their current location. SIP servers are co-located with SIP registrars. This enables the SIP server to find a user. |

Source: Camarillo

uses six methods of signaling, as shown in Table 5-1. Additional methods are under consideration by the IETF at this time.

When the call arrives at the remote endpoint, the phone rings and a new response is sent to that endpoint: RINGING (180). This is analogous to the Q.931 ALERTING message used in the PSTN. The time between the user dialing the last digit and the time RINGING is received by the caller is known as the *postdial delay* (PDD) for SIP call setup. If a telephone number is involved in addressing the call, the numbers must be translated into an IP address. Table 5-2 provides a comparison of SIP and PSTN signals.

When the called party answers the phone, a 200 response is sent to the calling party's UA. The UA sends another request, ACK, acknowledging the response to the INVITE request. At that moment, media begins to flow on the transport addresses of the two endpoints. The ACK may carry the final SDP parameters for the media type and format provided by the receiving

**Table 5-2**

Comparison of SIP
and PSTN signals

| SIP | PSTN |
| --- | --- |
| TRYING | Q.931 Call Proceeding |
| RINGING | Q.931 Alerting |
| ACK | Q.931 Connect |
| XINVITE | Q.931 Connect |

endpoint. The sequence of the INVITE and following ACK messages is similar to the Q.931 CONNECT message. ACKs do not require a response. Table 5-3 displays the response codes for SIP.

At this point in the call sequence, as seen in Figure 5-1, media flows over RTP, with the *Real-Time Control Protocol* (RTCP) providing the monitoring of the quality of the connection and its associated statistics. Next, as the name would imply, a BYE request from either party ends the call. As all messages are sent via UDP, no further action is required.

**SIP Calls via a Proxy Server** Proxy servers in the SIP sense are similar in function to proxy servers that serve a web site (mail relay via SMTP) for a corporate *local area network* (LAN). An SIP client in this case would send a request to the proxy server that would either handle it or pass it on to another proxy server that, after some translation, would take the call. The secondary servers would see the call as coming from the client. By virtue of receiving and sending requests, the proxy server is both a server and a client. Proxy servers function well in call forwarding and follow-me services.

Proxy servers can be classified by the amount of state information they store in a session. SIP defines three types of proxy servers: call stateful, stateful, and stateless. Call stateful proxies need to be informed of all the SIP transactions that occur during the session and are always in the path taken by SIP messages traveling between end users. These proxies store state information from the moment the session is established until the moment it ends. A stateful proxy is sometimes called a transaction stateful proxy as the transaction is its sole concern. A stateful proxy stores a state related to a given transaction until the transaction concludes. It does not need to be in the path taken by the SIP messages for subsequent transactions. Forking proxies are good examples of stateful proxies. Forking proxies are used when a proxy server tries more than one location for the user; that is, it "forks" the invitation.

**Table 5-3**

SIP response codes

| Number code | Description | Number code | Description |
|---|---|---|---|
| 100 | Trying | 413 | Request, entity too large |
| 180 | Ringing | 414 | Request, URI too large |
| 181 | Call is being forwarded | 415 | Unsupported media type |
| 182 | Queued | 420 | Bad extension |
| 183 | Session progress | 480 | Temporarily not available |
| 200 | OK | 481 | Call leg/transaction does not exist |
| 202 | Accepted | 482 | Loop detected |
| 300 | Multiple choices | 483 | Too many hops |
| 301 | Moved permanently | 484 | Address incomplete |
| 302 | Moved temporarily | 485 | Ambiguous |
| 305 | Use proxy | 486 | Busy here |
| 380 | Alternative service | 487 | Request cancelled |
| 400 | Bad request | 488 | Not acceptable here |
| 401 | Unauthorized | 500 | Internal severe error |
| 402 | Payment required | 501 | Not implemented |
| 403 | Forbidden | 502 | Bad gateway |
| 404 | Not found | 503 | Service unavailable |
| 405 | Method not allowed | 504 | Gateway timeout |
| 406 | Not acceptable | 505 | SIP version not supported |
| 407 | Proxy authentication | 600 | Busy everywhere |
| 408 | Request timeout | 603 | Decline |
| 409 | Conflict | 604 | Does not exist anywhere |
| 410 | Gone | 606 | Not acceptable |
| 411 | Length required | | |

Source: Camarillo

Stateless proxies keep with no state. They receive a request, forward it to the next hop, and immediately delete all states related to that request. When a stateless proxy receives a response, it determines routing based solely on the Via header and it does not maintain a state for it.[1]

An SIP call using a proxy server is a little more complicated than the simple SIP call model described previously (see Figure 5-3). In this call, the caller is configured with the called party's SIP server. An INVITE is sent to the called party's SIP server with the called party's text address in the To field. The called party's server determines if the called party is registered in that server. If the called party is registered on that server, it then determines the called party's current location on the network. This is called the mobility feature.

Once the called party is located via the mobility feature, the proxy server generates an INVITE request with no alteration of the headers of the request, except to add its own name in the Via field. Multiple servers may be involved in tracking down the called party.

Next the server must retain state information on the call. The server does this by correlating Cseq numbers, call IDs, and other elements of the headers as they pass through the proxy server. The server sends a TRYING message back to the calling party's agent. When the called party answers at the new location, a RINGING response is sent to the proxy server via the remote server (the called party's server). Both servers have Via entries in the response message to the calling party. Finally, ACK messages are

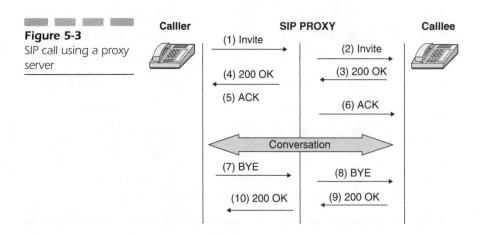

**Figure 5-3**

SIP call using a proxy server

[1]Camarillo, Gonzalo. *SIP Demystified*. New York: McGraw-Hill, 2002, p. 126–129.

exchanged, the call is established, and the media flow over RTP can begin. The call is terminated via a BYE request.

What makes the proxy server marketable is its user mobility feature. The called party can be logged in at multiple locations at once. This results in the proxy server generating the INVITE to all names on the list until the called party is found (preferably RINGING, but also TRYING or OK).[2]

A redirect server accepts SIP requests, maps the destination address to zero or more new addresses, and returns the translated address to the originator of the request. After that, the originator of the request can send requests to the addresses returned by the redirect server. The redirect server originates no requests of its own. Redirect servers pose an alternative means of call forwarding and follow-me services. What differentiates the redirect server from a proxy server is that the originating client redirects the call. The redirect server provides the intelligence to enable the originating client to perform this task as the redirect server is no longer involved.

The redirect server call model is a mix of the two previously described call models. Here a proxy model reverts to the direct call model once the called party is located. The redirect server returns a redirection response to the INVITE with code 301 or 302 indicating the called party is at the location listed in the Contact field of the message body. The calling party's *Media Gateway Controller* (MGC) closes its signaling with the redirect server and initiates another INVITE to the located returned in the redirect response. After that, the call flow is that of the direct model. If the called party is registered at a number of locations, the redirect server will return a list of names (URI) to be contacted. The calling party can then contact those addresses directly.

**Registrar** A registrar is a server that accepts SIP REGISTER requests. SIP includes the concept of user registration where a user tells the network that he is available at a given address. This registration occurs by issuing a REGISTER request by the user to the registrar. A registrar is often combined with a proxy or redirect server. Practical implementations often combine the UAC and UAS with registrars with either proxy servers or redirection servers. This can result in a network having only UAs and redirection or proxy servers.[3]

---

[2]Douskalis, Bill. *IP Telephony: The Integration of Robust VoIP Services*. New York: Prentice Hall, 2000, pg. 74.

[3]Collins, Daniel. *Carrier Grade Voice over IP*. New York: McGraw Hill, 2000, pp. 167–168.

**Location Servers**  Location servers are not SIP entities but are an important part of SIP architecture. A location server stores and returns possible locations for users. It can make use of information from registrars or from other databases. Most registrars upload location updates to a location server upon receipt. SIP is not used between location servers and SIP servers. Some location servers use the *Lightweight Directory Access Protocol* (LDAP, see IETF RFC 1777) to communicate with SIP servers.[4]

# New Standards for SIP

New standards efforts are aimed toward using the IP network for nonvoice interactions. Here are a few of the more exciting standards developments:

- *PSTN to Internet Interworking* **(PINT)**  The PINT (RFC 2848) standards effort defines "click-to-dial" services—interworking between a web page and a PSTN gateway element, or PINT server. PINT is helping to standardize service delivery, including request to call, request to fax, request to hear content, and, in the future, request to conference.

- *Services in the PSTN/IN Requesting Internet Services* **(SPIRITS)**  This working group is providing a framework for standardizing the mechanisms for controlling critical call altering and call-completion processes from the Internet by exposing the Internet domain to the PSTN's call model. The goal is to create a framework for Internet call-waiting-type services that will be applicable to wireline, wireless, and broadband (cable and *x-Type Digital Subscriber Line* [xDSL]) environments. Proposed SPIRITS-enabled network events include voice mail arrival, incoming call notification, attempts to dial numbers, dropping dialed connection, completing *Internet service provider* (ISP) connections, attempts to forward calls, and attempts to subscribe/unsubscribe to a PSTN service.

- *Signaling Translation* **(SIGTRAN)**  The IETF standard that deals with *Signaling System 7* (SS7; specifically *Transaction Capabilities Application Part* or TCAP over IP). A later chapter will cover SIGTRAN in greater detail.

- *Telephone Number Mapping* **(ENUM)**  The IETF standard that defines address mapping among phone numbers and Internet devices.

---

[4]Ibid., p.105.

It provides a way to reach multiple communication services via a single phone number. ENUM provides a process for converting a phone number into a DNS address (URL). It translates e.164 telephone numbers into Internet addresses.[5]

- ***Telephony Routing over IP* (TRIP)**   TRIP is a policy-driven interadministrative domain protocol for advertising the reachability of telephony destinations between location servers and for advertising attributes of the routes to those destinations. TRIP's operation is independent of any signaling protocol; hence, TRIP can serve as the telephony routing protocol for any signaling protocol.[6]

# Some SIP Configurations

SIP, given its simplicity and flexibility (discussed later in this chapter), simplifies the establishment of an alternative voice network. Figure 5-4 suggests a number of applications for SIP. The figure illustrates a corporate voice network that utilizes its corporate WAN to transport its interoffice voice traffic. This company can protect their investment in legacy phones and PBXs by installing an SIP gateway to interface the TDM technology with the SIP-based VoIP network. The SIP gateways also provide an interface with the PSTN.

In new offices and for remote workers, SIP phones provide a simple means of eliminating long-distance charges and connecting those workers to the rest of the network. Failing an expensive SIP phone, PC-based UAs such as those contained in Microsoft's XP (see the end of this chapter) can also function as a telephony agent connecting the worker to the network.

Providing the intelligence for this network is the SIP proxy server to control the calls. The application server provides critical corporate voice services such as voice mail, follow me, conferencing, call forwarding, unified messaging, voice-activated services, web-based provisioning, and new features that can be custom developed by the corporation, the SIP equipment vendor, or third-party vendors.

---

[5]Wolter, Charlotte. "ENUM Brings VoIP into Telephone Mainstream." *Sounding Board Magazine*, June 2001, p.12

[6]"Interworking Switched Circuit and Voice-over-IP Networks." A white paper by Performance Technologies and the International Engineering Consortium, available online at http://www.iec.org/online/tutorials/ip_in/topic05.html, pp. 5–6.

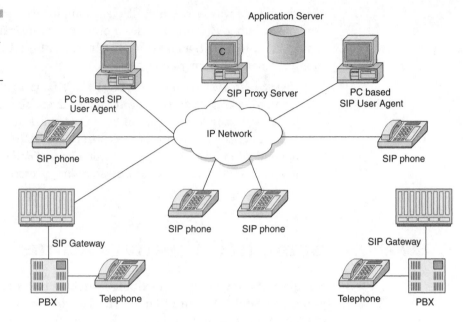

**Figure 5-4**
SIP gateways in long-distance bypass and enterprise telephony

# Comparison of SIP to H.323

H.323 is a blanket specification, meaning that it is not a protocol by itself, but rather defines how to use other protocols in a specific manner to create a service. H.323 was developed by the *International Telecommunications Union* (ITU) in the mid-1990s and consists of several protocols, including H.225 *registration, admission, and status* (RAS) signaling, H.225.0 call signaling, H.245 control signaling, and RTP. It also includes several standards for voice and video digitization and compression.

H.323 was originally designed to implement multimedia conferences on a LAN. In its original form, it was intended that certain conferences would be announced, or advertised in advance, and interested parties would subscribe or register to participate in the conference. These conferences may be interactive or may be broadcast events with participants viewing or listening only.

The applications were expanded to include participants who were not located on the same LAN as well as more general calling capabilities. This resulted in more complexity to the protocol and underlying applications. H.323 is a very large protocol suite, requiring a large amount of memory for code and data space. H.323 used a messaging scheme called *Abstract Syntax Notation* (ASN.1). ASN.1 is widely used by the ITU, but it is difficult for users to read and understand directly. The use of ASN.1 requires special test equipment to be built that can decode the ASN.1 messages and translate them into human-readable format for easier network testing and debugging.

The IETF sought to create a much simpler, yet powerful protocol that could be used for call setup and management in support of VoIP applications. A working group was formed, and the result of that effort was the development of the SIP, which was ratified by the IETF as RFC 2543 in 1999. H.323 and SIP have four fields of comparison: complexity, extensibility, scalability, and services.

# Complexity of H.323 Versus SIP

As VoIP protocols evolve, they become more efficient. Simplicity fosters acceptance. In this case, SIP is a marked improvement over H.323 primarily due to its greater simplicity, which translates to greater reliability. Given its simplicity relative to H.323, SIP enjoys a faster call setup, a prerequisite for carrier-grade voice applications.

H.323 is a rather complex protocol. The sum total of the base specifications alone is 736 pages. SIP, on the other hand, along with its call control extensions and SDPs totals 276 pages in RFC 3261. H.323 defines hundreds of elements, while SIP has only 37 headers (32 in the base specification, 5 in the call control extensions), each with a small number of values and parameters, but that contain more information.

A basic but interoperable SIP Internet telephony implementation can get by with four headers (To, From, Call-ID, and CSeq) and three request types (IN-VITE, ACK, and BYE) and is small enough to be assigned as a homework programming problem. A fully functional SIP client agent with a *graphical user interface* (GUI) has been implemented in just two man-months.

H.323 uses a binary representation for its messages, based on ASN.1 and the *packed encoding rules* (PER). SIP, on the other hand, encodes its messages as text, similar to HTTP[7] and the *Real-Time Streaming Protocol*

---

[7]Fielding, R., J. Gettys, J. Mogul, H. Nielsen, and T. Berners-Lee. Request for Comments (Proposed Standard) 2068: "Hypertext transfer protocol—HTTP/1.1." Internet Engineering Task Force, January 1997.

(RTSP).[8] This leads to simple parsing and generation, particularly when done with powerful text-processing languages such as Perl. The textual encoding also simplifies debugging, allowing for manual entry and the perusing of messages. Its similarity to HTTP also allows for code reuse; existing HTTP parsers can be quickly modified for SIP usage.[9]

One of the biggest criticisms of H.323 is that it can result in the transmission of many unnecessary messages across the network. This causes H.323 to not scale well. That is, for a large system, the overhead associated with handling a large number of messages has a significant effect on the system performance. A more heavily loaded system will result in poorer voice quality. QoS measures do not address network traffic and call setup issues in H.323.

H.323's complexity also stems from its use of several protocol components. There is no clean separation of these components; many services require interactions between several of them. (Call forward, for example, requires components of H.450, H.225.0, and H.245.) The use of several different protocols also complicates firewall traversal. Firewalls must act as application-level proxies,[10] parsing the entire message to arrive at the required fields. The operation is stateful since several messages are involved in call setup. SIP, on the other hand, uses a single request that contains all necessary information.

Figure 5-5 illustrates the messages required to set up a call, assuming that the caller already knows the address of the callee. Without going into the details of the messages, it can be seen that setting up the call requires the exchange of 16 messages across the network. This number grows considerably if the two terminals are located on different LANs or are managed by the Terminal Capability Set messages. This allows the two ends to agree on the parameters to be used for the call. In the case of SIP, this information is exchanged in the Invite and 200:OK messages (see Figure 5-6). Separate messages are not required. Likewise, in H.323, separate messages are exchanged to create logical connections between the two endpoints over which the voice information is passed (using the Open Logical Channel messages). In SIP, this information is also passed in the Invite and the 200:OK message.[11]

---

[8]Schulzrinne, H., R. Lanphier, and A. Rao. Request for Comments (Proposed Standard) 2326: "Real-time streaming protocol (RTSP)." Internet Engineering Task Force, April 1998.

[9]Schulzrenne, Henning, and Jonathan Rosenberg. "A Comparison of SIP and H.323 for Internet Telephony." White paper available at www.cs.columbia.edu/sip/papers.html.

[10]"Packet-Based Multimedia Communications Systems," H.323, ITU, February 1998.

[11]"SIP vs. H.323, A Business Analysis." White paper from Windriver, available at www. sipcenter.com/files/Wind_River_SIP_H323.pdf.

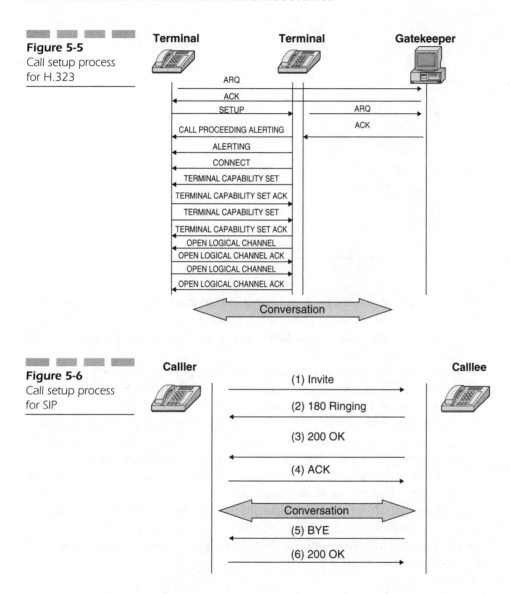

**Figure 5-5**
Call setup process
for H.323

**Figure 5-6**
Call setup process
for SIP

To counter this obvious weakness in H.323, procedures were added to enable H.323 to connect basic calls much quicker. This procedure is referred to as Fast Start and is illustrated in Figure 5-7. Fast Start eliminates several of the messages by requiring each side to have detailed knowledge of the parameters for the call and to have preassigned ports for communicating. This significantly limits the ability to make calls to multiple locations,

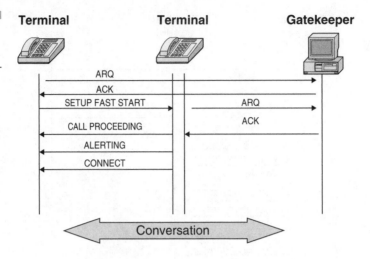

**Figure 5-7**
H.323 using
Fast Start

as the terminals must have configuration information for all locations they will contact using the Fast Start procedures. Furthermore, not all terminals support the Fast Start procedures, so this further limits the usefulness of this procedure.

Fast Start was developed assuming a fairly closed system, where the participants are limited and have intimate knowledge of each other, such as in an office environment where users are calling each other.[12]

An additional aspect of H.323's complexity is its duplication of some of the functionality present in other parts of the protocol. In particular, H.323 makes use of RTP and RTCP. RTCP has been engineered to provide various feedback and conference control functions in a manner that scales from two-party conferences to thousand-party broadcast sessions. H.245, however, provides its own mechanisms for both feedback and simple conference control (such as obtaining the list of conference participants). These H.245 mechanisms are redundant and have been engineered for small to medium-sized conferences only.[13]

## Scalability

H.323 and SIP differ in terms of scalability. The three main concerns are managing a large number of domains, server-processing capabilities, con-

---

[12]Ibid.
[13]Schulzrenne and Rosenberg.

ferencing, and feedback. H.323 was originally conceived for use on a single LAN and not a large number of domains. Issues such as wide area addressing and user location were not a concern. The newest version defines the concept of a zone and defines procedures for user locations across zones for email names. However, for large numbers of domains, and complex location operations, H.323 has scalability problems. It provides no easy way to perform loop detection in complex multidomain searches (it can be done statefully by storing messages, which is not scalable). SIP, however, uses a loop detection algorithm similar to the one used in BGP, which can be performed in a stateless manner.

Another factor for scalability is server processing. As regards server processing in an H.323 system, both telephony gateways and gatekeepers will be required to handle calls from a multitude of users. Similarly, SIP servers and gateways will need to handle many calls. For large, backbone IP telephony providers, the number of calls being handled by a large server can be significant. In SIP, a transaction through several servers and gateways can be either stateful or stateless. In the stateless model, a server receives a call request, performs some operation, forwards the request, and completely forgets about it. SIP messages contain sufficient state to enable the response to be forwarded correctly.[14]

Because the protocol is simpler, SIP requires less code to implement than H.323. This is reflected in lower fixed (code space) and dynamic memory requirements for both the protocol stack itself and the host application. Because of the smaller number of messages that the processor needs to handle to set up a call, SIP is faster than H.323. This means that SIP can process more calls per second than H.323 and that SIP requires a less powerful CPU to handle the same number of calls. The result is higher calls handled for a given system's usage. This results in larger message sizes for SIP than for H.323. However, it is better to send a few large frames of data than a lot of small frames except for slow links.[15]

SIP can be carried on either *Transmission Control Protocol* (TCP) or UDP. In the case of UDP, no connection state is required. This means that large, backbone servers can be based on UDP and operate in a stateless fashion, reducing significantly the memory requirements and improving scalability. H.323, on the other hand, requires gatekeepers (when they are in the call loop) to be stateful. They must keep the call state for the entire duration of a call. Furthermore, the connections are TCP based, which

---

[14]Ibid., pg. 3.
[15]"SIP vs. H.323, A Business Analysis."

means a gatekeeper must hold its TCP connections for the entire duration of a call. This can pose serious scalability problems for large gatekeepers.

Conferencing is another concern for scalability. H.323 supports multiparty conferences with multicast data distribution. However, it requires a central control point (called an MC) for processing all signaling, even for the smallest conferences. This presents several difficulties. Firstly, should the user providing the MC functionality leave the conference and exit their application, the entire conference terminates. In addition, since MC and gatekeeper functionality is optional, H.323 cannot support even three party conferences in some cases. We note that the MC is a bottleneck for larger conferences.

H.323 Version 2 has defined the concept of cascaded MCs, allowing for a very limited, application-layer, multicast distribution tree of control messaging. This improves scaling somewhat, but for even larger conferences, the H.332 protocol defines additional procedures. This means that three distinct mechanisms exist to support conferences of different sizes. SIP, however, scales to all different conference sizes. There is no requirement for a central MC; conference coordination is fully distributed. This improves scalability and complexity. Furthermore, as it can use UDP as well as TCP, SIP supports native multicast signaling, enabling a single protocol to scale from sessions with two to millions of members. Table 5-4 compares the SIP and H.323 call features.

Feedback is another concern when comparing H.323 and SIP. H.245 defines procedures that enable receivers to control media encodings, trans-

**Table 5-4**

Comparison of SIP and H.323 call features

| Feature | SIP | H.323 |
| --- | --- | --- |
| Blind transfer | Yes | Yes |
| Operator-assisted transfer | Yes | No |
| Hold | Yes, through SDP | No |
| Multicast conferences | Yes | Yes |
| Multiunicast conferences | Yes | Yes |
| Bridged conferences | Yes | Yes |
| Forward | Yes | Yes |
| Call park | Yes | No |
| Directed call pickup | Yes | No |

Source: Schulzrenne and Rosenberg

mission rates, and error recovery. This kind of feedback makes sense in point-to-point scenarios, but it ceases to be functional in multipoint conferencing. SIP, instead, relies on RTCP for providing feedback on reception quality (and also for obtaining group membership lists). RTCP, like SIP, operates in a fully distributed fashion. The feedback it provides automatically scales from a two-person point-to-point conference to huge broadcast-style conferences with millions of participants.[16]

## Extensibility

Extensibility is a key metric for measuring an IP telephony signaling protocol. Telephony is a tremendously popular, critical service, and Internet telephony is poised to supplant the existing circuit-switched infrastructure developed to support it. As with any heavily used service, the features provided evolve over time as new applications are developed. This makes compatibility among versions a complex issue. As the Internet is an open, distributed, and evolving entity, one can expect extensions to IP telephony protocols to be widespread and uncoordinated.

This makes it critical to include powerful extension mechanisms from the outset. SIP has learned the lessons of HTTP and SMTP (both of which are widely used protocols that have evolved over time) and built in a rich set of extensibility and compatibility functions. By default, unknown headers and values are ignored. Using the Require header, clients can indicate named feature sets that the server must understand. When a request arrives at a server, it checks the list of named features in the Requires header. If any of them are not supported, the server returns an error code and lists the set of features it does understand. The client can then determine the problematic feature and fall back to a simpler operation.

To further enhance extensibility, numerical error codes are hierarchically organized, as in HTTP. Six basic classes exist, each of which is identified by the hundreds digit in the response code (refer to Table 5-3). Basic protocol operation is dictated solely by the class, and terminals need only understand the class of the response. The other digits provide additional information, usually useful but not critical. This allows for additional features to be added by defining semantics for the error codes in a class while achieving compatibility. The textual encoding means that header fields are self-describing.

---

[16]Schulzrenne and Rosenberg.

As new header fields are added in various different implementations, developers in other vendors can determine usage just from the name and add support for the field. This kind of distributed, documentation-less standardization has been common in the SMTP, which has evolved tremendously over the years. As SIP is similar to HTTP, mechanisms being developed for HTTP extensibility can also be used in SIP. Among these are the *Protocol Extensions Protocol* (PEP), which contains pointers to the documentation for various features within the HTTP messages themselves.

H.323 provides extensibility mechanisms with some limitations. First, extensions are limited only to those places where a nonstandard parameter has been added. If a vendor wants to add a new value to some existing parameter, and no placeholder exists for a nonstandard element, one cannot be added. Secondly, H.323 has no mechanisms for enabling terminals to exchange information about which extensions each supports. As the values in nonstandard parameters are not self-describing, this limits interoperability among terminals from different manufacturers.

H.323 requires full backwards compatibility from each version to the next. As various features come and go, the size of the encodings will only increase. However, SIP enables older headers and features to gradually disappear as they are no longer needed, keeping the protocol and its encoding clean and concise. A critical issue for extensibility are audio and video *coder/decoders* (codecs). Hundreds of codecs have been developed, many of which are proprietary. SIP uses the SDP to convey the codecs supported by an endpoint in a session. This is due to the differences of ASN.1 versus text.

This means that SIP can work with any codec, and other implementations can determine the name of the codec and contact information for it. In H.323, each codec must be centrally registered and standardized. SIP enables new services to be defined through a few powerful third-party call-control mechanisms. These mechanisms enable a third party to instruct another entity to create and destroy calls to other entities. As the controlled party executes the instructions, status messages are passed back to the controller. This allows the controller to take further actions based on some local program execution. SIP enables these services to be deployed by basing them on simple, standardized mechanisms. These mechanisms can be used to construct a variety of services, including blind transfers, operator-assisted transfers, three-party calling, bridged calling, dial-in bridging, multiunicast to multicast transitions, ad hoc bridge invitations and transitions, and various forwarding variations.[17]

---

[17]Schulzrinne, Henning, and Jonathan Rosenberg. "Signaling for Internet Telephony." Technical Report CUCS-005-98. New York: Columbia University, February 1998.

Another aspect of extensibility is modularity. Internet telephony requires a large number of different functions; these include basic signaling, conference control, QoS, directory access, service discovery, and so on. One can be certain that mechanisms for accomplishing these functions will evolve over time (especially with regards to QoS). This makes it critical to apportion these functions to separate modular, orthogonal components, which can be swapped in and out over time. It is also critical to use separate, general protocols for each of these functions.

H.323 is less modular than SIP. It defines a vertically integrated protocol suite for a single application. The mix of services provided by the H.323 components encompass capability exchanges, conference control, maintenance operations, basic signaling, QoS, registration, and service discovery. Furthermore, these are intertwined within the various subprotocols within H.323. SIP's modularity enables it to be used in conjunction with H.323. A user can use SIP to locate another user, taking advantage of its rich multihop search facilities. When the user is finally located, he or she can use a redirect response to an H.323 URL, indicating that the actual communication should take place with H.323.[18] Table 5-5 compares H.323's call control service with that of SIP.

**Table 5-5**

*Call control service comparison H.323 to SIP*

| Criteria | H.323 | SIP |
|---|---|---|
| Complexity | Very complex | Simple |
| Message set | Many messages | Few messages |
| Debugging | Have to alter tools when a protocol is extended | Simple tools |
| Extensibility | Extensible | Very extensible |
| User extendable | ASN.1—complex | Text based—easy |
| Elements that must maintain states | Clients, gatekeepers, MCU, gateways, UAs, some proxy servers | |
| Processor usage | More overhead | Less overhead |
| Telephony features | Robust | Robust |
| Host application | Very complex | Much simpler |
| Code size | Large | Small |
| Dynamic memory usage | Large | Small to medium |

[18]Ibid.

## Services

What makes SIP and softswitch architectures so revolutionary is the ease with which they offer services and features. At the minimum, SIP supports personal mobility via its use of registration such that a corporate user can move his or her SIP phone anywhere on the corporate LAN or WAN, and the network will instantly know where to route that user's calls. One-number service is made possible by SIP forking proxies.

SIP messages carry *Multipurpose Internet Mail Extension* (MIME) content in addition to an SDP description. A response to an INVITE message could include a piece of text, an HTML document, an image, and so on. As an SIP address, a URL can be included in web content for click-to-call applications. SIP was designed for IP-based applications and can leverage those applications to create new telephony-oriented services.[19]

Almost as exciting as new services is the fact that SIP can be used to implement existing *Custom Local Area Signaling Service* (CLASS) services found in the PSTN (for example call forwarding, conferencing, caller ID, and so on). One significant objection service providers have had about VoIP and softswitch is the notion that softswitch architectures did not replicate the legendary 3,500 features that came with the 5ESS CLASS 5 switch. Given the flexibility of SIP and *voice XML* (VXML) software, those 3,500 features, in theory, can be matched or exceeded. This will be addressed in a later chapter comparing softswitch to CLASS 4 and 5 switches.

Signaling for SIP calls is routed via a proxy. This enables the proxy to utilize advanced feature logic. The feature logic may be resident at the proxy or in a separate feature server or database. The proxy may have access to other functions such as the policy server, an authentication server, and so on, each of which may be distributed throughout an IP network or all in one location. For example, the policy server may hold call-routing information or QoS-related information while the feature server holds subscriber information. The SIP proxy server can interface with SS7 signaling at the *service control point* (SCP) where services are offered via the *Intelligent Network Application Part* (INAP).[20] Interfacing SS7 and IP networks is addressed in a later chapter.

Given that SIP is a text-based language, many programmers in the industry are familiar with web applications, text parsers, and scripting languages that contribute to SIP. No special knowledge is necessary to create

---

[19]"SIP vs. H.323, A Business Analysis."
[20]Collins, pg. 207–208.

an SIP service. Also, any PC can be an SIP server to test new applications. SIP enables programmers familiar with the functional needs of a certain community to create their own services. Gone are the days when a service provider has to wait 18 months from the request to the delivery of a single new feature.

## H.323 Versus SIP Conclusion

If SIP is better, why is H.323 important? A large, installed base of H.323 equipment exists, and vendors building new VoIP products are concerned about being backwards compatible with existing VoIP equipment and services. Carriers building VoIP networks also want to be able to enable their users to communicate with users on other VoIP networks. This would imply that H.323 is still an important protocol. Based on the technical advantages of SIP, the widespread support of SIP by H.323 vendors, the large number of SIP vendors already in the market, and the nearly universal availability of SIP services from the service providers, there is little argument in favor of continuing to develop H.323-based products.

# The Big "So What?" about SIP

SIP offers some immediate impact on the telecommunications industry. Given its simlicity and ease of use, it has aleady found a numer of applications that could catapult its use into the mainstream in just a few years.

## SIP on Windows XP

In March 2001, Microsoft announced that its new version of Windows, Windows XP, would contain SIP. This has immense implications. If every PC running Windows XP is an SIP UA, then every one of those computers is potentially a telephone and is at least somewhat independent of the PSTN. This would be especially true of PCs on corporate WANs where PC-to-PC calls would become the norm rather than the exception. For users with anthropological barriers to using a PC as a phone, very inexpensive handsets are available that interface the PC to the user.

Microsoft projects upwards of 245 million XP users by the end of 2002. If, eventually, all 245 million XP users shifted at least some of their telephony usage to their PC via Windows XP SIP applications, there would be some significant loss of traffic to the legacy telephone companies. No telephone company in the world has 245 million subscribers. What if Windows XP SIP becomes to telephony what Microsoft Word has become to word processing?

## How Does That Work?

Microsoft, by developing and distributing Windows XP with SIP, has facilitated some deconstruction of the telecom industry as we know it. To make an SIP call via Windows XP, the user gains access to the Internet (potentially via Microsoft Network, Microsoft's ISP) or their IP corporate WAN. It should be noted here that the MSN ISP has some 5 million subscribers. With the acquisition of a few large U.S. ISPs, Microsoft Network could be one of the largest if not the largest ISP in the United States.

Using their account with one of the partner IP telephony service providers (Net2Phone, DialPad, or Delta3), users can place a call to any phone in the world. WorldCom plans to offer VoIP services to XP users in the enterprise market. In the United Kingdom, Callserve Communications plans to do the same in the residential market. In Canada, Microsoft has partnered with Telus, offering voice-over-Internet services to high-speed subscribers.

Referring back to the PSTN model comprised of access, switching, and transport, Microsoft assumes the position of providing access to the voice network. Specifically, the Windows-XP-powered PC potentially replaces the handset; assuming the subscriber is using Microsoft Network as its ISP, Microsoft is the de facto local loop. Switching is provided by the IP telephony carriers (DialPad, Net2Phone, and so on) and transport is via an IP backbone service provider (Level3 or Cable and Wireless).[21] As a result, the central office is potentially replaced.

The specific agent in Windows XP that contains the SIP package is Windows Messenger. To deliver Internet telephony to the Windows user, Windows Messenger offers three broad technological improvements. First, it significantly reduces voice delay, the chief quality issue with VoIP, to a min-

---

[21]Arnold, Jon. "Where Does Microsoft Want To Go Today?" *Sounding Board Magazine*, December 2001. www.soundingboardmag.com/articles/1c1sb.html.

imum of about 70 milliseconds. When communicating clients are on one well-engineered LAN, this delay is not noticeable. When clients communicate over the Internet, there is substantial headroom (at least 80 milliseconds) before delays begin to interfere with conversational dynamics.

Second, Windows XP comes with a variety of voice codecs that Windows Messenger uses. When network conditions favor wide bandwidth and low delay, these make voice sound significantly better than telephone quality. When network throughput degrades or delay increases, coders are able to gracefully fall back to lower bit rates. When network congestion eases again, coders dynamically step up to higher quality. This happens automatically, under application control.

Third, Windows Messenger includes acoustic echo cancellation software, which reduces echo and eliminates the need for a headset in average-sized offices. Windows Messenger doesn't solve Internet delay or reachability problems, but it appears at the right time. Today the Internet is significantly faster than it was five or six years ago. Many, many more people have PCs. Increasingly, these PCs are always on. In the ideal workplace, these PCs are usually connected to the Internet via 100 Mbps Ethernet. At home, residential always-on connections (DSL, cable, and wireless) are becoming more and more prevalent. Increasingly, PCs can place IP phone calls to any other phone in the world using gateway providers, as mentioned previously.[22]

This product may find a very fertile market outside the United States, particularly in nations where even local calls (which is most of the world outside the United States) are billed by the minute and it takes years to get a landline phone installed (assuming alternatives to the last mile). Assuming a similar proportion of long-distance business worldwide is comparable to the 70 percent figure in the United States, non-U.S. businesses will want to move their interoffice phone traffic, both local and long distance, on to their corporate WAN courtesy of the SIP package contained in Windows XP. For many in the developing world, Internet access is gained at a local cyber café. PCs in those cyber cafés could be configured to perform as telephones. This will have the effect of introducing competition *to* the local loop in those countries.

---

[22]Isenberg, David. "Windows Messenger: New Waves of Innovation." A white paper for Microsoft, www.microsoft.com/windowsxp/pro/techinfo/planning/networking/windowsmessenger.asp.

# Conclusion

This chapter discussed SIP as its own softswitch architecture. Rather than be a compromise between TDM and IP networks, SIP was designed from the start as a signaling protocol for VoIP. Its architecture is more scalable, extensible, and simple. The consensus in the industry is that SIP poses a better value proposition for service providers and corporate networks than does H.323. The prospect of offering services via an SIP infrastructure that cannot be provided by TDM switches has many service providers considering SIP as an easy-to-deploy VoIP architecture solution.

It should be noted here that in 1995 Microsoft introduced its Windows 95 product containing networking protocol TCP/IP. Within approximately two years, TCP/IP became the de facto networking protocol for enterprise networks as almost all corporate PCs were loaded with and networked with Windows 95. H.323 also came into prominence via its inclusion in Microsoft's NetMeeting product. Given these precedences, a strong possibility exists that the inclusion of SIP in Windows XP, due to the sheer volume of the number of PCs that will ultimately be loaded with Windows XP, could well establish SIP as the dominant signaling protocol for VoIP.

# Softswitch: More Scalable Than CLASS 4 or 5

# Scalability

One of the most promising aspects of softswitch architecture is that its flexibility in scaling makes so many revolutionary applications possible. Given this flexibility in scaling, it is now possible to bypass the central office. A *point of presence* (POP) for a long-distance company can now consist of a four-port media gateway. As addressed in Chapter 2, "The Public Switched Telephone Network (PSTN)," the Class 4 switch scales to approximately 100,000 ports. For mainstream service providers, the term "scalability" is an interrogative for "how big" does the product in question scale to. They have traditionally thought in terms of large, centrally located, and managed switches. A large platform often translates into a lower-cost-per-port expense for the service provider. Scalability for a softswitch is contingent on two elements: the total number of ports and the call-processing capabilities. The softswitch industry, and the media gateway industry in particular, is only recently introducing high-density gateways that compete with the Class 4 and 5 switch in terms of port density (see Figure 6-1).

Scalability as regards the total number of ports is contingent upon the media gateway. In 1998, the author sold media gateways that totaled 168 ports per platform. This was the most densely populated product of its kind at the time. "Racking and stacking" these would not prove difficult, but

**Figure 6-1**
The PSTN and the relationship of Class 4 and 5 switches with centralized Class 4 switches

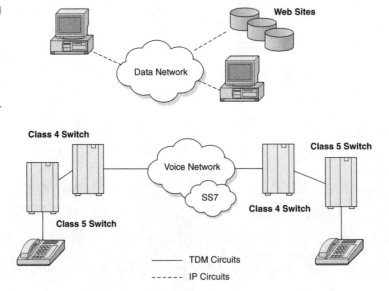

managing them would. Also, the real estate costs of having so many low-density gateways in a central office or other facility would prove prohibitively expensive. Some of today's gateways offer densities in excess of 100,000 DS0s per 7-foot rack. Some media gateways can be clustered into super nodes to achieve densities in the low hundreds of thousands of DS0s. Table 6-1 illustrates the progress in media gateway density.

In the converging market, scalability in terms of high density in gateways is not a major issue. Because service providers use IP networks, which can be connected to diverse locations in North America, a high-density, centralized architecture for which the Class 4 or 5 switch was designed is not necessary. Rather, less dense media gateways distributed on an IP network are more advantageous in the converging market. This enables a service provider to enter or exit a market with less risk than they would encounter with a Class 4 or 5 switch. A Net Present Value analysis demonstrates this advantage in financial terms.

# Scaling Up

Two quantifiable components of scalability exist in reference to softswitch. They are the number of ports (DS0s) supported and the call-processing power. If these softswitch components are to overcome the objections of service providers, softswitch solution vendors must increase the density of their media gateways and boost the call-processing power of their softswitch solutions.

Since 1998, the industry has progressed in the direction of ever-denser media gateways and efforts to manage "super nodes" where media gateways can be clustered to offer upper-end scalability comparable to that of the Class 4 or 5 switch. Considering that Netrix Corporation offered one of the densest gateways on the market in 1998 with 168 DS0s, it is clear that media gateways can exceed Class 4 or 5 in scalability per 7-foot rack.

**Table 6-1**

Media gateways offer greater density in one 7-foot rack than Class 4

| Media gateway | Density (DS0s/7-foot rack) |
|---|---|
| Cisco MGX8260 | 16,000 |
| Convergent Networks ISC2000 | 109,000 |
| Sonus GSX9000 | 24,000 |

Softswitch is smaller than Class 4 or 5 in its footprint (smaller size and shape). As illustrated in Figure 6-2, softswitch requires less physical space to deliver the same port density, that is, the equivalent number of phone lines, as Class 4 or 5. Depending on the configuration, a softswitch may take as little as one-thirteenth of the space required by Class 4 or 5 to perform the same scale of service. This is a powerful advantage for softswitch. Due to the smaller footprint, softswitch power consumption and cooling requirements are also less than the legacy switches. Smaller hardware size also translates into lower "real estate" expenses.

One comparison is a softswitch solution from Convergent Networks. Their product can concentrate over 100,000 DS0s in one 7-foot rack consuming a 24-inch × 24-inch space, with space to access the switch, for a billable total of 12 square feet per month. A Class 4 in comparison delivering 24,000 DS0s requires 13 racks for a total of 156 square feet at $35 per square foot of central office space per month. The softswitch in this example offers a 92 percent cost saving over the Class 4 in real estate costs.[1]

The miniaturization of Class 4 components will not soon match the advantages of softswitch in footprint. The smaller footprint of the softswitch translates into lower real estate costs for the service provider. Although this

**Figure 6-2**
Footprint comparison of Class 4 or 5 switches and a high-density Voice over IP (VoIP) media gateway

Billable space necessary to house 13 racks (36,000 DS0s) for Class 4 – 156 sq.ft.

Space necessary to house 1 rack (36,000 DS0s) for Softswitch - 12 sq. ft

[1]Convergent Networks. "A Business Case for Next-Generation Switching: Revolutionizing Carrier Revenues and Returns." www.convergentnet.com.

may not be a major concern for an incumbent provider who has built and paid off its central office facilities, it is critical for a new market entrant or an incumbent expanding into new markets. Figure 6-2 illustrates the financial advantages in rent or real estate concerns relative to the smaller footprint of softswitch.

Scaling up a media gateway is a factor of hardware and software. A high-density design consists of call-processing modules, PSTN interface modules, and packet interface modules. The packet interface modules provide the interface to the packet-switched network. Packet interfaces include DS-3, OC-3 optical interfaces, multiple 100 BaseT, and Gigabit Ethernet interfaces. DS-3 (672 DS0s)/OC-3 capacities are relatively recent improvements in the upward scalability of media gateways. Table 6-1 and Figure 6-3 illustrate the progression of media gateways (and softswitch solutions) in port density.

The switch fabric module of a media gateway performs the routing of packets through the system. Line cards fill out the appropriate header

**Figure 6-3**
Softswitch is overcoming the performance deficiency of low density relative to Class 4.

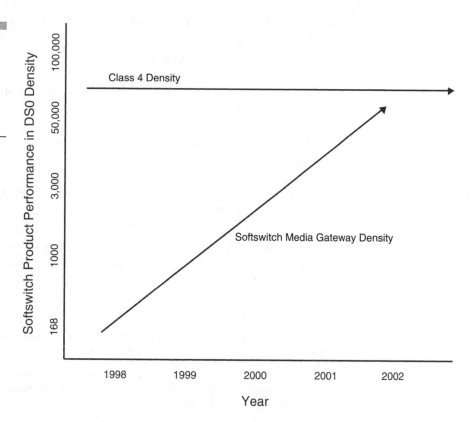

information that is used by the switch fabric to direct packets to the appropriate line card/external interface. The VoIP modules consist of a "farm" of *digital signal processors* (DSPs) that perform the actual conversion of the voice streams between the PSTN and packet worlds. The DSP industry has strived to increase the ratio of conversations per DSP chip to help the media gateway industry achieve greater gateway density. Progressions in DSP chip technology and the software that controls those chips enable greater media gateway density.[2]

In summary, by building denser media gateways, softswitch vendors have overcome the objection that the softswitch solution will not scale up to a port density comparable to a Class 4 or 5 switch (see Figure 6-3). This is a matter largely of mechanical and software engineering to build a bigger gateway. Figures 6-3 and 6-4 demonstrate how softswitch compares favorably with Class 4 or 5 in scalability both in terms of density and *busy-hour call attempt* (BHCA) processing.

In the United States, approximately 1,400 Class 4 switches and 25,000 Class 5 switches are being utilized. An alternative long-distance solution, for example, would be to install a media gateway(s) in each central office and connect them via an IP network. As a result, the market demand would be for media gateways that scaled down rather than up. If that were the case, softswitch, given its flexibility to scale up as well as down, would be in greater demand than Class 4.

The other issue of scalability is the call-processing power of softswitch. The *media gateway controller* (MGC) in the softswitch solution handles this. Early versions of softswitch could not compete with Class 4, for example, in terms of calls per second or BHCAs. Recent releases from softswitch vendors offer BHCA counts well into the millions (such as Clarent at 5 million). This exceeds Nortel's DMS-250, which boasts 700,000[3] and competes with Lucent's 4ESS at a stated 5 million BHCAs.[4] In 2000, softswitches were reported to offer BHCA ranges from 250,000 to 500,000.[5] Figure 6-4 illustrates this progression in BHCA counts. It should be noted that no independent verification of these BHCA counts exists either for legacy vendors of Class 4 and 5 switches or for softswitch solutions.

---

[2]Texas Instruments. "Carrier-Class, High Density Voice over Packet (VoP) Gateways." A white paper hosted by the International Engineering Consortium, 2001, www.iec.org.
[3]Nortel Networks. "The DMS300/250 System Advantage." Product brochure at www.nortel.com.
[4]Lucent Technologies. "Long Distance Network Solutions." Product brochure at www.lucent.com.
[5]Wolter, Charlotte. "Long and Winding Road." *xchange Magazine.* July 15, 2000. www.x-changemag.com/articles/072feat2.html.

**Figure 6-4**

Softswitch call-processing capacity now exceeds that of the Nortel DMS-250 Class 4 switch (Source: Sounding Board Magazine).

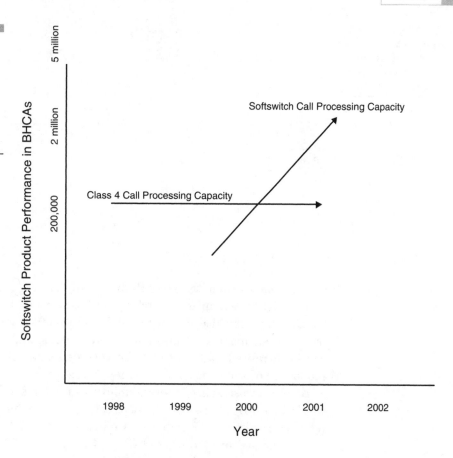

The limited number of Class 5 replacements taking place at the time of this writing (mid-2002) usually consists of port-for-port replacements of the Class 5 by a dense media gateway at the central office. The softswitch and application servers are usually colocated with the media gateways. As this technology gains experience in the marketplace, examples of distributed architecture will be more clearly in evidence. This chapter will address alternative architectures later on.

## Scaling Down

Although most of the early objections to VoIP and softswitch solutions held by service providers regarded scaling as high as a Class 4 or 5 switch, much of that debate failed to focus on the flexibility offered by VoIP solutions in

**Figure 6-5**
An IP phone as the
least dense network
configuration

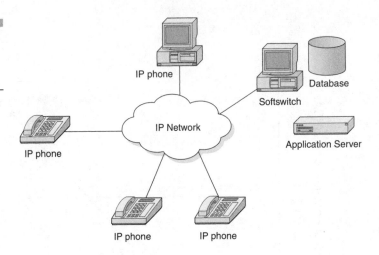

scaling *down* rather than *up*. Rather than focusing a network on a very dense, highly centralized switching architecture with its incumbent and astronomical expense, softswitch architecture offers far greater flexibility (read cost savings) over Class 4 or 5 switches. Originally, telephone switches were centralized and needed to be very dense to achieve the maximum economies of scale and to save money on real estate and maintenance.

At the lowest end of the scalability range for softswitch architecture would be an IP phone, which constitutes the equivalent of *one* DS0 on a telephone switch (see Figure 6-5). Contrast that with the minimum configuration for a Nortel DMS-250 Class 4 switch at 480 DS0s. The service provider does not need to invest in a switch to service a customer with an IP phone. The service provider provides only the IP connection or, at a minimum, the proxy and application server to complete the call over the IP network (that could include the public Internet). A data service provider provides that connection (a *data local exchange carrier* [DLEC] or an *application service provider* [ASP]).

## Scaling Down for Class 4 Applications

Another revolutionary approach of interim architecture is the inclusion of VoIP gateways as replacement technology to the Class 4 switch for long-distance applications. Figure 6-6 details this solution. Refer back to Figure 6-1 detailing the PSTN and the architecture for Class 4 and 5 networks. Rather than route traffic to a Class 4 switch, the traffic goes to a VoIP gateway for packetization and routing over an IP network.

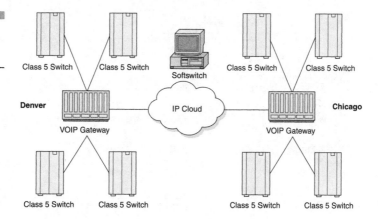

**Figure 6-6**
VoIP gateway as
Class 4 replacement

This architecture could be taken one step further with the inclusion of a VoIP gateway colocated at every Class 5 switch (see Figure 6-7). For many service providers, this would pose the advantage of eliminating backhaul or "hairpins," that is, the interswitch transportation on expensive circuit-switched networks. This configuration could prove especially useful in linking rural or outlying communities by using less dense VoIP gateways (*small office/home office* [SOHO] or enterprise grade) described in Chapter 3, "Softswitch Architecture or 'It's the Architecture, Stupid!'"

In these settings, the central office may consist of a *private branch exchange* (PBX). Such a PBX configured with an internal or external VoIP gateway could prove more cost effective for the service provider than the highly centralized, legacy *Time Division Multiplexing* (TDM) structure. This would be especially valuable for an *Internet service provider* (ISP) that wanted to leverage their investment in their existing infrastructure and customer base by offering long-distance services. This is accomplished by the installation of a low-density media gateway at the central office and the trunk side of their gateway connecting to their existing IP connection. Refer to Figure 6-6 for details of this architecture.

## Scaling Down for Class 5 or Central Office Bypass Applications

Scaling *down* rather than up becomes truly revolutionary in the Class 5 bypass, also known as Class 5 replacement applications (see Figure 6-8). Referring back to Figure 6-5, in an architecture where all subscribers have an IP phone, no switch or media gateway is necessary. This has the

**Figure 6-7**
Colocating VoIP
gateways at each
Class 5 switch

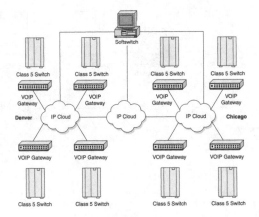

**Figure 6-8**
IP phones and
softswitch as a Class
5 replacement

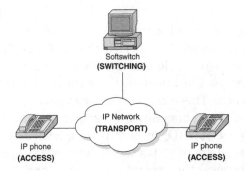

potential to make a Class 5 switch obsolete (but not soon). This network would be comprised of IP phones and their data connections as the form of access. A softswitch or a *Session Initiation Protocol* (SIP) application performs the switching. Transport is in the form of the IP network, including the Internet. This scenario assumes what is a highly futuristic setting at the time of this writing (mid-2002), given that the access portion for over 90 percent of all residences, for example, are controlled by *Regional Bell Operating Companies* (RBOCs) via a central-office-centric architecture. A form of broadband IP or Internet connection would make this possible via a *digital subscriber line* (DSL), cable modem, or a fiber-to-the-home connection to the residence.

## Scaling Down for Class 5 or Central Office Bypass from the Residence

Another advantage of scaling down rather than up comes in the form of a residential media gateway, as described in Chapter 3. A residential media gateway consists of one to four ports for an analog telephone (RJ-11 connection). The residential gateway, if it does not have one integrated, connects to a residential router in the case of a DSL application or set-top box in the case of a cable modem. Some cable modems have a residential media gateway built in. Once the voice stream is packetized in the residential media gateway, it is largely independent of Class 4 and 5 switches, unless the call is terminated on the PSTN (see Figure 6-9).

The next step up in scalability regarding a Class 5 replacement or bypass would consist of aggregating residential lines at a media gateway located near the residence. The simplest scenario would be the installation of a media gateway in a pedestal found at most suburban street corners. Rather than aggregate multiple copper wires into a T1 or T3 for transmission to the central office, the media gateway would aggregate those lines into an IP connection (physically an Ethernet connection), packetize the voice stream, and connect with the IP network. Because the switching function is performed by a softswitch over a distributed architecture network, it is *not* necessary to route the call via a Class 5 switch for a local call or via a Class 4 switch for a long-distance call. In this instance, the softswitch can be described as a classless switch, because geography is meaningless. It is not necessary to break out the call as being completed within a certain geography that in the past was limited by copper wire and Class 4 or 5 switching.

It should be noted here that not any media gateway will work for this application. Such a media gateway would have to be environmentally hardened to withstand the stresses of high and low temperatures. It would also have to have a chip set engineered for the more demanding rigors of a Class

**Figure 6-9**
Residential media
gateway as a Class 5
bypass

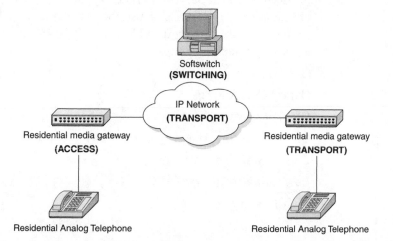

5 application. One such demand is the need to recognize *dual-tone multi-frequency* (DTMF) tones, which is not present in all media gateways.

## Access Switching

The access switching application provides a solution for next-generation end-office switching, using a *Next-Generation Digital Loop Carrier* (NGDLC) in the role of the access media gateway. The access (softswitch) switch acts as the *Signaling System 7* (SS7) *service switching point* (SSP), terminating SS7 signaling to and from the PSTN. It also provides call processing, billing, services, management, and signaling services to the service provider, leveraging existing installed NGDLC equipment for bearer resources. From a PSTN network perspective, the access (softswitch) switch performs the role of an end-office switch.

The following end-office capabilities would be provided by an access (softswitch) switching application:

- *ISDN User Part* (ISUP) bearer trunks from the NGDLC
- *Channel Associated Signaling* (CAS) trunks with *multifrequency* (MF) address signaling from NGDLC
- Operator services *end-office* (EO) signaling
- E911 call routing
- *Communication Assistance for Law Enforcement Act* (CALEA)
- 1-8XX toll-free routing
- End-office *local number portability* (LNP)
- Subscriber-line supervision and call processing with NGDLC *Plain Old Telephone Service* (POTS) subscriber linecards (no GR303 or TR008 switch connection is required for these subscribers), including all necessary telephony tones
- Call Waiting
- Cancel Call Waiting
- Call Forwarding, No Answer
- Call Forwarding, Busy
- Call Forwarding, Variable
- *Basic business groups* (BBG)
- Multiline hunt groups

- Voice mail, including Visual Message Waiting Indication and *SCSI Musical Data Interchange* (SMDI) connectivity
- Caller line identification (name and number)
- Call Trace
- Three-way conferencing
- Call transfer
- Tones and announcements
- Local tandeming

This scaling down, made possible by access switching, could also prove very valuable to alternative service providers such as cable TV companies, power companies, municipal utility companies (power and water) and *wireless ISPs* (WISPs).

What makes a traditional telephone company? As described in Chapters 2 and 3, it is three elements: access, switching, and transport. These three types of entities all provide access for the customer (business or residence) *and* a means of transport on their networks. Traditionally, the body of knowledge and capital necessary to deploy, operate, and maintain Class 4 and 5 switches was too great for any entity but a telephone company to maintain. Assuming the customer maintains their own media gateway in the form of IP phones or low-density media gateways, a service provider is freed of the expense of legacy switching. Their switching would be performed by a softswitch hosted on an industrial-grade server located anywhere on their network. By selling voice services as a value added to their cable TV, power and water, or ISP services, these entities will present serious competition to incumbent telephone companies. Figure 6-10 shows an example of a Class 5 bypass media gateway.

**Figure 6-10**

*Class 5 bypass media gateway at a residential pedestal, providing access switching*

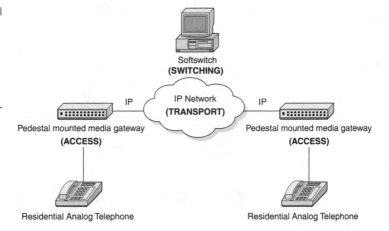

Softswitch
**(SWITCHING)**

IP — IP Network **(TRANSPORT)** — IP

Pedestal mounted media gateway
**(ACCESS)**

Pedestal mounted media gateway
**(ACCESS)**

Residential Analog Telephone

Residential Analog Telephone

## Scaling Down for Class 5 or Central Office Bypass from the Enterprise

Some 70 percent of corporate long distance is interoffice long distance, that is, employees of one company calling one another long distance. Enterprise IP telephony bypasses Class 4 and 5 switches. Enterprise media gateways, when used in conjunction with an IP PBX, constitute a central office bypass from what historically has been the most lucrative market-business voice service. Corporate long distance is the cream of the long-distance business. If 70 percent of corporate long distance migrated off the networks of the incumbent long distance service providers and on to corporate *wide area networks* (WANs), many of those service providers would be severely disrupted.

The capability for a media gateway to scale down rather than up makes the migration of corporate telephony over a softswitch architecture possible (see Figure 6-11). IP PBX and IP Centrex are examples of softswitch architectures on corporate WANs.

Scaling down media gateways also makes the IP Centrex based on a Class 5 possible. One geographically independent Class 5 has the potential to service any telephone handset (analog or IP) anywhere on the network. The network can also be accessed via the Internet. Figure 6-12 details how scaling down frees the corporate voice network from dependence on a legacy long-distance provider and even local service providers. An example of this architecture is an IP Centrex product from AG Communications called iMerge.

**Figure 6-11**

Corporate softswitch solutions that bypass Class 4 and 5 made possible by scaling down the media gateway

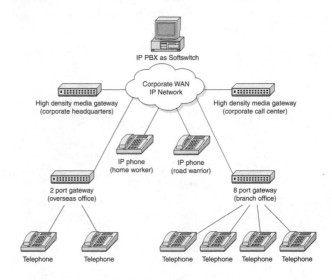

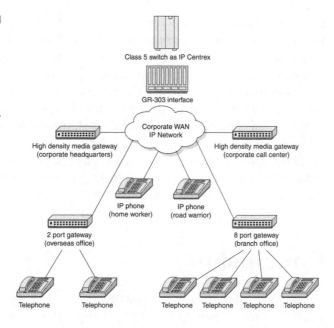

**Figure 6-12**
IP Centrex using one Class 5 is made possible by scaling down the media gateway.

## Scaling Down Technical Issues

Some media gateway vendors have focused on ever-denser gateway switches, emulating the centralized architecture of the circuit-switched industry. What happens when a service provider must provide service in a low-density market (a small city, town, or suburb)? This presents a problem when those high-density media gateways must interoperate with low-density gateways (residential or enterprise media gateways). The issues that prevent interoperability include the following:

- The dissimilar gateways may not share the same VoIP signaling protocol (H.323, SIP, or *Media Gateway Control Protocol* [MGCP]).
- Even where the gateways share the same signaling protocol, they may be of different vendors, which often results in incompatibility.

As the scenarios described previously in this chapter suggest, there is no "one size fits all" approach for a softswtiched network. A number of vendors offer a very dense gateway packaged with a softswitch to pursue the carrier-grade market. This works for applications when it directly replaces a Class 4 or 5 switch, which is exactly what many have done to date with those high-density media gateways. A complication arises when high-density media gateways must interoperate with low-density media gateways of dissimilar vendors and dissimilar signaling protocols (SIP versus H.323) installed at remote locations that require only a very low density media gateway (four or eight ports, for example). Far-sighted vendors of dense media gateways have partnered with vendors of low-density media gateways with extensive interoperability testing to ensure they can provide potential customers with a flexible range of media gateway densities.

No vendor, not even those that claim to have end-to-end solutions, makes a line of gateways from low density (4 ports, for example) to carrier grade (100,000 ports) that is fully interoperable. It is then necessary to use a softswitch to translate between signaling protocols and dissimilar vendor platforms in order to optimize flexibility in scaling.

## Conclusion

The scenarios described here involving media gateways that emphasize the value of scaling down rather than up support an emerging picture of the virtual telephone company. A virtual telephone company owns no Class 4 or 5 switches. It maintains a softswitch, an application server(s), and a billing platform. It completes calls over an IP network maintained by an IP backbone service provider. The virtual telephone company provides call control over the IP network via its softswitch. It provides services such as conferencing, call forwarding, voice mail, and so on from its application servers. Some companies are emerging that exist to provide special calling services (an example would be illuminets that provide signaling services). It should also be noted that legacy service providers that offer data services, such as AT&T and Sprint, can also benefit when their subscribers (usually corporate) take advantage of media gateways that scale down rather than up. Figure 6-3 and 6-4 provide a comparison of the port density of Class 4 switches and Class 4 replacement softswitches. In summary, softswitch solutions are just as, if not more, dense than Class 4 (Nortel's DMS-250 in this case).

According to Clayton M. Christensen in *The Innovator's Dilemma,* disruptive technology is defined as "cheaper, simpler, smaller, and more convenient." Such a definition can be used to describe a media gateway compared to a Class 4 or 5 switch. Scaling down certainly constitutes being smaller than a Class 4 or 5 switch. With IP phones coming on the market at less than $100 per handset and 4-port media gateways also at about $100, they are cheaper per port than Class 4 or 5. It becomes even cheaper for the service provider when the customer is purchasing the media gateway for his or her own use and spares the service provider of having to make this investment. Assuming the customer is buying and maintaining the IP phone or media gateway, it becomes infinitely simpler for the service provider in that the customer is maintaining the IP phone or media gateway. The service provider must merely maintain the softswitched network.

Scaling down potentially disaggregates the PSTN for the quasi-monopolistic service providers. The Internet has disaggregated a number of industries over the last few years. Examples include travel agencies, mortgage brokers, and car dealers. Telephone companies may be next on the list. Delivering voice services is no longer contingent upon being large enough to deal with the economy of scale that many argue was necessary in order to offer voice service. Entities smaller than a multibillion dollar RBOC can offer voice by taking advantage of low-density (and low-cost) media gateways working with a softswitch to provide call control and features.

An example would be an ISP that installed media gateways in the calling areas it served and routed long-distance voice services via its IP services. Community communications cooperatives could install media gateways in curbside pedestals, such as those used by the incumbent phone service. This would enable subscribers their choice of the incumbent telco service via the legacy TDM infrastructure or the community coop service via the media gateway installed on the curb. This is called access switching and is a maturing subset within the softswitch industry.

The scaling-down model will be most valuable in the developing world where the discretionary incomes of the majority of the population cannot support the installation of expensive, high-density Class 4 and 5 switches. Telephone service can be extended to remote villages in the form of IP phones (one IP phone for a small village, for example) or low-density gateways (four to eight ports, for example). Given the low cost of installation and the operation of low-density applications, service providers, local co-ops, or even governments can offer telephone service at prices affordable to the target populations.

Finally, the last point of Christensen's definition of disruptive technology is met in that IP phones and media gateways are "more convenient to use" for the service provider if the customer is buying and maintaining them. The service provider is freed from the onerous expense of maintaining high-density switches like the DMS-250 Class 4 switch, which comes with a $500,000 per year maintenance contract or a Class 5 switch at upwards of $2 million per year (depending on the size). Service providers freed of these expenses can undercut incumbent service providers who must continue to maintain a legacy network of Class 4 or 5 switches. Such competition would ignite a "race to the bottom" in service pricing, greatly benefiting consumers and proving ruinous to incumbents operating under a cumbersome, expensive legacy paradigm. In this way, softswitch solutions that incorporate low-density media gateways disrupt legacy Class 4 and 5 switch vendors as well as legacy telephone service providers.

# Softswitch Is Just as Reliable as Class 4/5 Switches

A recurring objection to *Voice over IP* (VoIP) and softswitch solutions is the perception that such a solution would not match the "five 9s" of reliability provided by the Class 4 and 5 switches. This chapter will set forth that a softswitch solution is just as reliable (actually "available") as the Class 4 or 5 switch, and a softswitched network is potentially more reliable than the PSTN. Much of what applies to building highly reliable carrier-grade softswitch solutions can work for enterprise-grade networks as well.

# World Trade Center Attack: A Need to Redefine Reliability

The September 11th, 2001 attack on the World Trade Center has served to focus attention on the vulnerabilities of the legacy, circuit-switched telephone network. Verizon, the largest telephone company, had five central offices serving some 500,000 telephone lines south of 14th Street in Lower Manhattan. More than six million private circuits and data lines passed through switching centers in or near the WTC. AT&T and Sprint switching centers in the WTC were destroyed in the attack. Verizon lost two WTC-specific switches in the towers, and two nearby central offices were knocked out by debris, fire, and water damage. Cingular Wireless lost six and Sprint PCS lost four. Power failures interrupted service at many other wireless facilities (see Figures 7-1 and 7-2).[1]

Verizon further estimates 300,000 voice business lines, 3.6 million data circuits, and 10 cellular towers were destroyed or disrupted by the events of September 11th. This equates to phone and communications service interruption for 20,000 residential customers and 14,000 businesses.[2] The question a subscriber inevitably has to ask is "Where are the five 9s of reliability in this system?" What is wrong with the PSTN that one incident can deprive so many of service?

The American *Public Switched Telephone Network* (PSTN) can be described as having a centralized architecture. The telephone companies have not built redundancy into their networks. Almost all cities and towns across the nation rely on one hub or central office, meaning that if that hub

---

[1]Telecom Update #300, September 17, 2001. www.angustel.ca/update/up300.html.
[2]Naraine, Ryan. "Verizon Says WTC Attacks May Hurt Bottom Line." *Silicon Alley News*. www.atnewyork.com/news/article/0,1471,8471_897461,00.html.

**Figure 7-1**
Class 5 switch destroyed in WTC attack

**Figure 7-2**
Central offices and other switching facilities were concentrated in the WTC and 140 West Street. What if every building in Manhattan had its own media gateway on a softswitched, distributed network?

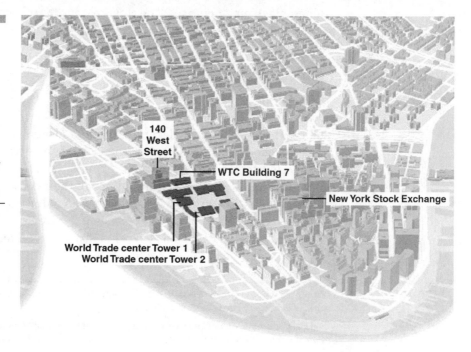

were to be destroyed, that city would lose all land-line telephone connectivity with the outside world. Even with the growth of *competitive local exchange carriers* (CLECs), fewer than 10 percent of those CLECs have facilities truly separate from the *Regional Bell Operating Companies* (RBOCs). Between 1990 and 1999, the number of RBOC central offices grew less than 1 percent to a nationwide total of 9,968, while the total number of phone lines grew by 34 percent according to the *Federal Communications Commission* (FCC).[3]

The FCC's *Automatic Reporting and Management Information System* (ARMIS) tracks failures caused by telephony switches, which in turn generates availability figures. The FCC is not unmindful of the threat posed to the nation's voice telephony infrastructure and the FCC has announced the creation of a Homeland Security Policy Council. The Council's missions are to assist the FCC in evaluating and strengthening measures for protecting U.S. communications services, and to ensure the rapid restoration of communications services and facilities that have been disrupted as the result of threats to, or actions against, our nation's homeland security. The Council also ensures that public safety, health, and other emergency and defense personnel have effective communications available to them to assist the public as needed.[4]

Without a drastic change in the architecture of the PSTN, it will remain vulnerable to major outages. The chief reasons for this are not that the *network elements* (NEs) in the form of Class 4 or Class 5 switches are less than reliable, but rather the architecture centers on *single points of failure* (SPOF): the central offices and the star network architecture that comprises the PSTN.

One of the myths that supports the survival of Class switches and their architecture is the notion that only a Class 4 or 5 switch has five 9s of reliability. The perception in legacy telecommunications circles is that Bell Labs somehow received divine providence for the design of a switch that would consistently deliver five 9s of reliability and no other platform could mimic this unassailable standard. Although the switches themselves might be very reliable, the architecture may contain vulnerabilities.

This chapter will explore what is meant by five 9s and why engineering a network to deliver that level of reliability is only a matter of good engi-

---

[3]Young, Shawn, and Dennis Berman. "Trade Center Attack Shows Vulnerability of Telecom Network." *Wall Street Journal*, October 19, 2001.

[4]Federal Communications Commission. "Federal Communications Commission Announces Creation of Homeland Security Policy Council." Press release. November 15, 2001. www.fcc.gov.

neering not limited to Bell Labs designs. Five 9s applies to data networks as well as to telephone switches. Data networks have long been engineered to achieve five 9s. VoIP is voice over a data network. Ergo a VoIP network can be engineered to deliver five 9s of reliability. A VoIP network can be just as, if not more, reliable than the PSTN.

## One to Five 9s

As described in Table 2-4 in Chapter 2, "The Public Switched Telephone Network (PSTN)," downtime in a network can be very expensive. Uptime is referred to as availability. Availability is often expressed numerically, as a percentage of uninterrupted productive time containing from one to five 9s. For instance, 99 percent availability, or two 9s, equates to a certain amount of availability versus downtime, as does 99.9 percent (three 9s), and so on.

Key contributors to reduced availability include planned site mainte-nance, operator errors, site-wide disasters, planned database reconfigura-tion, planned OS patches, and networking errors.

## Standards for Availability

Telcordia, formerly Bellcore, has compiled a number of specifications that define the standards for reliability in network components for both circuit and packet-switched technologies. It should be noted here that no one agency monitors a switch or other network element for its level of avail-ability. Independent auditors can be hired to perform testing for availabil-ity along the lines of the specifications established in Table 7-1.

# What Is Reliability?

The terms reliability and availability are often used interchangeably, but they are two distinct measures of quality. Reliability refers to component failure rates measured over time, usually a year. Common reliability mea-sures of components are the annual failure rate, *failures in time* (FIT), *mean time between failure* (MTBF), *mean time to repair* (MTTR), and SPOF. Table 7-2 outlines the availability terms.

**Table 7-1**

Telcordia specifications related to system reliability

| Telcordia specification | Description |
| --- | --- |
| GR-512 | Defines downtime and reporting requirements for downtime. Focuses on *Integrated Services Digital Network* (ISDN) and covers topics such as redundancy, failover, and so on. |
| GR-529 | Defines FCC outage reporting criteria for carriers. |
| GR-499 | Defines inter-*Local Access and Transport Area* (LATA) transport availability objectives and *automatic protection switching* (APS) switchover requirements, including specific time boundaries for when switchovers should occur. |
| TR-TSY-511 | Defines service standards and provides definitions for what constitutes a cutoff call and an ineffective machine attempt. |
| GR-1110 | Defines reliability metrics of circuit-switched networks for packet-switched networks. Follows the PSTN reference model and represents the key elements of GR-512, GR-499, and TR-TSY-511. |
| SR-3580 | Includes the *Network Equipment Building Standards* (NEBS) that address the physical reliability of a switch. |
| GR-454 CORE | Entitled "General Requirements for Supplier Provided Documentation" and provides information on developing complete and comprehensive documentation. |
| GR-2914 CORE | Entitled "Human Factor Requirements for Equipment to Improve Network Reliability" and focuses on designing equipment from a user's perspective. |

Source: Convergent Networks and ATIS

**Table 7-2**

Terms related to availability

| Term | Definition |
| --- | --- |
| Annual failure rate | The amount of downtime expressed as the relationship between the MTBF and the number of hours in a year (8760). |
| FIT | FIT is the total number of failures of a module in a billion hours (1,000,000,000 hours). |
| MTBF | Mean Time Between Failures—the average time a manufacturer estimates before a failure occurs in a component or complete system. As MTBF is an estimate of averages, half of the devices can be expected to fail before that time and the other half after. |
| MTTR | An estimate on the part of the vendor as to the average time necessary to do repairs on equipment. |
| SPOF | A single point or network element at which failure could bring down a network or subnetwork. |

# How Availability Is Calculated

Availability indicates the system uptime from an operation perspective. System availability is a function of aggregate component reliability; thus, availability is likewise measured in terms of time.

**Calculating Availability and Unavailability**   The availability of a hardware/software module can be obtained by the following formula where A is availability.

$$A = MTBF/MTBF + MTTR$$

In the following formula, U is unavailability:

$$U = MTTR/MTBF + MTTR$$

$$Availability = 1 - Unavailability$$

The product of the formula gives the percentage of "up time" for the system, that is the amount of time for which it is "available." This is how availability is expressed. The target for a telephone network is 99.999 percent "up time" availability. This is known as "five 9s" of eliability.

**Annual Failure Rate**   The hours per year are 8760.

$$AFR = 8760/MTBF$$

The greatest requirement for availability is for NEs, such as a circuit switch, which are generally required to provide 99.999 percent availability or 0.001 percent unavailability.[5] Such a system is termed *Highly Available* (HA).

The previous equations apply to each individual component on a network. How is availability determined for a network that has hundreds or thousands of components?

Figure 7-3 illustrates how different components on a network may have different levels of downtime or unavailability. In order to determine the availability of a network, it is necessary to add up the availability of all components on a network.

---

[5]Convergent Networks. "Understanding Carrier-grade Reliability and Availability." www.convergentnetworks.com.

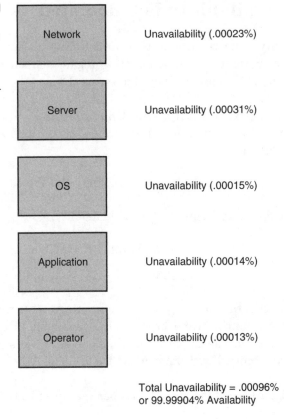

**Figure 7-3**
System availability is a function of the availability of all the components.

Network — Unavailability (.00023%)

Server — Unavailability (.00031%)

OS — Unavailability (.00015%)

Application — Unavailability (.00014%)

Operator — Unavailability (.00013%)

Total Unavailability = .00096%
or 99.99904% Availability

## How Does a Switch, PSTN or Softswitch, Achieve Five 9s?

Selling against the reputation of the Class 4 for reliability is the most difficult problem for softswitch vendors. "Where are the five 9s of reliability?" is the inevitable question service providers pose to vendors. Although a number of softswitch vendors claim to achieve five 9s, much skepticism exists among service providers that the five 9s of the new technology have the same experience as a Class 4. If the "heavy iron" of a Class 4 switch can achieve five 9s, then how does a software-based product like softswitch achieve five 9s in reliability? A service provider's confidence is further challenged by softswitch platforms that are not NEBs compliant. Given the cost of downtime, as discussed earlier in this chapter, long-distance service providers, for example, will insist that the components on their network provide five 9s.

Achieving five 9s is mostly a matter of engineering. If softswitch vendors can engineer out SPOFs and engineer in redundancy and other measures that Class 4 and 5 switch vendors have used for years to ensure reliability, then they, like the Class 4 and 5 switch vendors, can also advertise five 9s of reliability. Softswitch vendors can also engineer their platforms to be NEBS compliant. These measures enable softswitch to match five 9s of reliability demonstrated by Class 4. A number of softswitch solutions have achieved five 9s.

HA is simply a matter of good engineering. The main components of engineering a Class 4 or 5 switch for HA are redundancy, no SPOF, a hot switchover, a preservation of calls, in-service upgrades, component reliability, reproducible quality, and NEBS compliance. The same is true for a softswitched network.

HA is enhanced when each component is replicated in a system. This is called redundancy. If one unit fails, its replicated unit takes over. Redundant configurations are expressed by the notation $m:n$, where $m$ represents the number of standby units and $n$ represents the number of active units supported by the standby units. A typical configuration is 1:1 where there is one active unit for every active unit, or 1:6 where there is one standby unit for 6 active units. Usually, the smaller the $n$, the greater the protection and cost. Given the highly reliable nature of today's components, a carrier may determine that configurations greater than 1:1 provide sufficient availability. In a PSTN configuration, 1:$n$ redundancy schemes are employed on the access side (a Class 5 switch, for example) where an outage affects a smaller number of subscribers. Class 4 and 5 switches are more likely to use a 1:1 redundancy model because the effect of a failure is more expensive. The effect of Moore's Law, where computing power doubles while computing cost halves every 18 months, makes redundancy less expensive as time goes by.[6]

In an HA system, two or more systems are loosely coupled together with each other, with the help of redundancy software. The reliability provided can be further classified as asymmetric or symmetric based on whether the systems act as active/standby (idle) or run in a parallel load-sharing/balancing mode. In an Active/Standby type of system, further categories exist, such as 1+1 redundancy or N+K redundancy based on the number of active nodes and standby nodes that are available. Cluster mode is another

---

[6]Ibid.

such HA architecture where applications can run in either load-sharing or failover mode. The reliability of HA systems can be further enhanced by hardening some of the hardware components of the individual system constituting an HA system. Typical candidates for such treatment are the *network interface card* (NIC), disk controller, disk, power supply, and so on.[7]

HA computing utilizes the redundant resources of clustered (two or more) processors. Such solutions address redundancy for all the components of a system, processors, main and secondary memory, network interfaces, and power/cooling. Although hardware redundancy may be effectively addressed by clustered (redundant) hardware resources, the class of errors detected is less comprehensive and the time required to recover from errors is much longer than that of fault-tolerant machines. Still, fault recovery in tens of seconds, from most common equipment failures, can be achieved at less than half the cost of traditional fault-tolerant computer systems. High-availability systems are often configured as dual-redundant pairs.[8]

Carriers require high system availability and are concerned with the effects of possible softswitch downtime. Carriers demand low MTBF and employ traffic overload control, the shedding of call-processing capacity in the event of component failures, and quick failure detection and recovery mechanisms. The softswitch answer is to architect redundant softswitch hardware nodes at different locations throughout the network, which contributes to the overall network reliability. The majority of softswitches on the market run on servers from either Sun or Hewlett Packard.

Recent advances in technology and declines in computer component pricing show the rate of Moore's Law may be somewhat less than 18 months. Moore's Law states that computing power (the power of semiconductors) doubles every 18 to 24 months, while the price of that computing power decreased 54.1 percent each year from 1985 to 1996. At the same time, the Gross Domestic Product deflator increased by 2.6 percent per year. Communications equipment, in this case both Class 4 and softswitch, is based largely on semiconductors. Hence, Moore's Law applies to both softswitch and Class 4 and 5, but more so to softswitch, given its client/server architecture (a newer platform than the Class 4 and 5 mainframe). The explostion in DS0 density in media gateways and BHCAs possible in softswitch exceeds Moore's law (computing power doubles every eighteen months). The count of DS0's per 7-foot rack grew from 1,176 for a Netrix

---

[7]Ibid.

[8]Kehret, William. "High Availability Requires Redundancy and Fault Tolerance." *RTC Magazine*, Volume VI, Number 2, February 1998. www.themis.com/new/pubs/ha_article.html.

media gateway in 1998 to 108,864 for a Convergent Networks media gateway in 2001, marking a growth in capacity of 30 times per year. In addition, Dale W. Jorgenson of *American Economic Review* has stated that software investment is growing rapidly and is now much more important than investment in computer hardware. Given that the main cost of a softswitch solution is software, this is very true of softswitch. If these trends continue, it is wise to use two years as a life cycle for these projects.[9] This makes systems of six or more 9s economically feasible.

Another mechanism to enhance reliability is to ensure that no SPOF is on the system. That is, every mechanism has a backup in the event of the failure of one unit. Hot standby refers to having a replicated unit take over the functions of its primary unit for either planned or unplanned outages. Continuously available (reliable) systems rely exclusively on active replication to achieve transparency in masking both planned and unplanned outages.[10] In the PSTN, which relies on a star network topology and central offices, a central office or single fiber optic cable can represent an SPOF (see Figure 7-4).

A crucial element of a high availability system design is to not only have redundant components, but to be able to switch over to the redundant components without a loss of voice or data traffic (see Figure 7-5). The architecture and implementation of hot switchover has one of the most significant effects on overall system availability in running live traffic. Elements of hot switchover include rapid execution and rapid fault detection. Rapid execution is the capability of a system to perform hot switchover to the standby mode in a subsecond timeframe. Rapid fault detection is the time required to detect a fault. This should range from microseconds for readily detectable events to full seconds for difficult-to-detect faults.

The preservation of calls and billing is the ultimate goal of a highly available voice system in the event of a switchover. In this case, the user detects no disruption of the call, and the call is properly billed with no loss of revenue to the service provider. Technically, this means a preservation of the voice bearer traffic and the successful reestablishment and synchronization of all associated signaling.

In-service upgrades are an important aspect of any service provider network. The ability to upgrade a voice system while it is simultaneously carrying live traffic is elemental for carriers. Live upgrades concern two factors:

---

[9]Jorgenson, Dale W. "Information Technology and the U.S. Economy." *American Economic Review*, Vol. 91, 2001. pp. 3–7.

[10]Clustra. "The five 9's Pilgrimage: Toward the Next Generation of High Availability Systems." www.clustra.com. pg. 3.

**Figure 7-4**
This damaged RBOC pedestal constitutes an SPOF on the network. Note 4×4s holding the structure up and plastic trash bags providing weather resistance.

**Figure 7-5**
HA systems replicate all network elements, resulting in no SPOF. (Source: Hewlett-Packard)

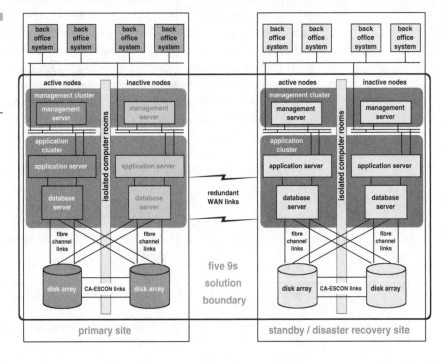

software and hardware. A sound hot switchover design goes a long way toward making both of these possible. To update existing and running software while simultaneously running live traffic, it is necessary to switch over to the new software with no impact to service. To update hardware without service interruption, a "hot swap" is performed on the circuit packs. That is, one circuit pack is removed from the system without a negative effect on service.

What Class 4 and Class 5 aficionados are referring to when they speak of five 9s is component reliability. System availability is a function of the reliability of the individual components. The best way to ensure a highly available system is to avoid failure in any component of a system. This is best accomplished using only the best quality parts with minimal failure rates.

Network connectivity for a softswitch product includes interfaces to the *Internet Protocol* (IP) world and interfaces to the PSTN world. PSTN world connectivity is primarily through DS0 or higher types of interfaces, which are point to point and cannot be inexpensively replicated. Generally, for signaling, several parallel channels operate between the two network points through separate hardware interfaces and whenever one channel fails, traffic is diverted on to the parallel channels. Primarily, failing NICs or a lost network connection causes IP world connectivity failures.

Floating IP and Floating *Media Access Control* (MAC), along with multiple points of connectivity to the IP cloud, are some of the concepts used for guarding against network failures. But problems with maintaining *Transmission Control Protocol* (TCP)/IP states across platforms require special handling in software. It is easier to handle failovers based on the *User Datagram Protocol* (UDP).

Database failures can be caused by disk failures, disk controller failures, database corruption, and so on. Duplicating the hardware involved and replicating the database across multiple points protect against such database failures. Environmental and physical failures are difficult to predict and prevent. These include causes like fire and earthquakes, which can be protected against by geographical redundancy. ("Geographical redundancy" refers to locating softswitch components at various locations around the country or around the world. In the event of a failure of one switch in New York City, for example, a back-up switch in Omaha, Nebraska, for example, would pick up the load of the New York-based failed switch.) It should be noted that since these systems work in parallel, a software bug in one system would also get replicated on the other side and lead to downtime.[11]

---

[11]Hughes Software Corporation. "Challenges in Building Carrier Grade Convergence Products." March 2001, pg. 7. This white paper is available online at www.hssworld.com/whitepapers/voppapers.htm#5

Finally, reproducible quality dictates that strict operational controls are required to ensure that a supplier has predictable results with respect to the reliability of the systems produced. Such controls range from parts acquisition, components testing, system assembly, and the method of installation. A failure in operations can effectively eradicate the benefits provided by an elegant system design.[12]

## Network Equipment Building Standards (NEBS)

In addition to five 9s, the other buzzword for reliability in the Class 4 market is NEBS. NEBS addresses the physical reliability of a switch. It is contained in Telcordia specification SR 3580, an extensive set of rigid performance, quality, safety, and environmental requirements applicable to network equipment installed in a carrier's central office. Most carriers in North America require that equipment in their central offices be NEBS compliant.

NEBS testing includes electrical safety, immunity from electromagnetic emissions, lightning and power faulting, and bonding and grounding evaluations. Equipment must achieve standards including temperature, humidity, and altitude testing; fire resistance (usually by destructive burning); earthquake vibration resistance; and other rigid tests. NEBS compliance includes ensuring access to mirror sites, fire- and waterproof storage facilities for database and configuration backup information, and backing up electrical power in case of power failures.

With five 9s and NEBS, the Class 4 and 5 vendors have developed, through decades of experience, a very reliable product that delivers superb uptime and rapid recovery capability. As a result, service providers are very reluctant to experiment with new technologies (see Figure 7-6).

# Specifications for Softswitch Reliability

No readily accepted industry standards exist for softswitch design and performance. As mentioned in Table 7-3, a Telcordia specification is available for packet-switched systems: GR-1110-CORE. It maps the reliability met-

---

[12]Convergent Networks.

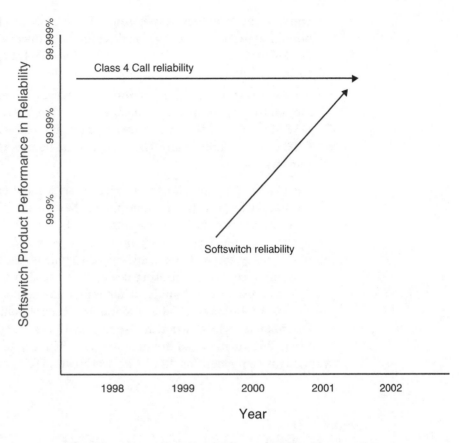

**Figure 7-6**
Softswitch now
meets the reliability of
a Class 4 switch
(Source: Data sheets
from softswitch
vendors).

**Table 7-3**

GR-1110 down-
time specifications

| GR-1110-CORE reliability parameter | Downtime (minutes/year) |
| --- | --- |
| Core system downtime | 0.4 |
| Individual port interface downtime | 12 |
| Multi-interface downtime | 1.2 |

rics defined in the circuit-switch specifications and applies them to the packet-switched industry. Key measures include the following:

■ **Total service downtime**   The expected long-term average annual time spent in failure mode (due to hardware failures and *operations, administration, and maintenance* [OA&M] activities) that affects all services.

- **Individual interface downtime**   The expected long-term average annual time spent in failure modes (due to hardware failures and OA&M activities) that affects one interface (DS-1, DS-3, or OC-12, and so on). The allowed time is 12 minutes per year.

- **Multiport interface downtime**   The expected long-term average annual time spent in failure modes (due to hardware failures and OA&M activities) that affects two or more interfaces on a card (DS-1, DS-3 or OC12, and so on). The specification defines this time to be 1.2 minutes per year.

GR-1110 is ATM-centric and makes no specific mention of IP. The Packet-Cable standards provide a reliability model for VoIP with a cable slant.

By using many of the mechanisms that Class 4 and 5 switches have utilized over the years (redundancy, fault tolerance, and NEBS) to achieve five 9s of reliability, softswitch is achieving the same levels of reliability. Given the declining costs of computing power, it is possible that softswitch may even exceed five 9s of reliability while remaining economically competitive to a Class 4 solution (see Table 7-4 and note the reliability figures.).

Distributed architecture can also improve the reliability of a softswitch solution. With distributed architecture, no SPOF exists on a network. Any redundant component on the IP network can pick up where the primary

**Table 7-4**

Comparing softswitch products to Class 4 switches

| Vendor and product | DS0s/ rack | BHCAS | Reliability | NEBS3 | MOS | Price per DS0 |
|---|---|---|---|---|---|---|
| Nortel DMS-250 (Class 4) | 2,688 | 800,000 | 99.999 | Y | 4.0 | $100 |
| Lucent 4ESS (Class 4) | 2,688 | 700,000 | 99.999 | Y | 4.0 | $100 |
| Convergent ICS2000 (Softswitch) | 108,864 | 1,500,000 | 99.9994 | Y | 4.0 | $25 |
| SONUS GSX9000 (Softswitch) | 24,000 | 2,000,000 | 99.999 | Y | 4.0 | $25 |
| Nuera Nu-Tandem (Softswitch) | 6120 | 480,000 | 99.999 | Y | 4.83 | $75 |

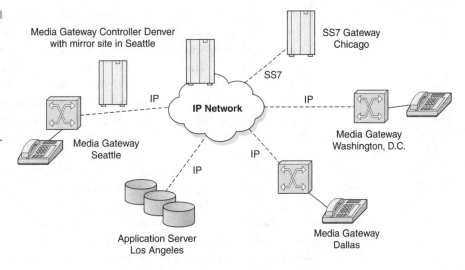

**Figure 7-7**
Distributed architecture provides greater survivability/reliability over the centralized architecture of the PSTN.

component failed. Figure 7-7 illustrates a dispersal of softswitch components around the United States. If a *media gateway controller* (MGC) in Denver is destroyed in a force majeure, another media gateway controller can pick up where the Denver MGC failed.

# Software Reliability Case Study Cisco IOS

Software failures are perhaps the most difficult sources of failures to get out of any system. As software gets more complex and hardware more reliable, software-related outages might become more frequent, though more sophisticated software development, testing, and debugging techniques would try to counterbalance this effect. A typical strategy is to improve software testing and provide software resilience with a redundant set of software running on duplicated hardware. The important point is that the redundant software should not be processing the same inputs as the failing software at any point in time. In the case of convergence products, protection against software/hardware failures can be categorized as warm or hot, depending on whether any of the ongoing calls are preserved after failover (the process of switching from failing system to running system) or not.

To accurately predict the availability of a network or device, the software contribution to downtime must be considered. The problem, however, is that no industry standard exists for software reliability. Several methods are available for calculating software reliability, but these methods are known to be inaccurate and difficult to perform. The following paragraphs detail a study of one software product and are not intended as an endorsement of any product line.

The best method known is to analyze field measurements for software reliability. As an example, Cisco Systems examined the number of software-forced crashes from large sample sets with a variety of Cisco IOS (a Cisco software package) versions. Cisco standardized on a six-minute MTTR as a conservative estimate for the time it would take a router to reboot and reestablish neighbors and traffic flows in a medium-sized enterprise environment. These measurements did not account for software defects that did not restart the device, as these were difficult to measure. Using this methodology, Cisco concluded that Cisco IOS *General Deployment* (GD) software meets the five 9s reliability. Cisco IOS Release 11.1 is 99.99986 percent reliable with an MTBF of 71,740 hours and that the non-GD version of Release 11.1 is 99.9996 percent reliable with an MTBF of 25,484 hours.

Software reliability is primarily a factor of software maturity. As the software ages and gains acceptance, defects are reduced until the software is considered stable with no problems expected. Developing the appropriate release can take 18 months. Cisco is also committed to releasing five 9s software by conducting extensive release tests on new software.[13]

## Power Availability

Power and environment are also potential influences on the overall availability of the telephone network. Power is also unique in that it does not impact one device at a time like software or hardware. It affects, or can affect, an entire building or multiple buildings at a time. This can impact all available devices on the network, including distribution, the core, the gateway, and softswitch all at once. The calculations therefore change from

---

[13]Cisco Systems. "IP Telephony: The Five Nines Story." p. 12, available at www.cisco.com/warp/public/cc/so/neso/vvda/iptl/5nine_wp.htm

device based to entire network based, which creates a significant impact on the theoretical availability depending on the power protection strategy used for the network.

## Typical Power Outages in a Typical Telephone Network

What does a power outage look like in a Central Office? What is its impact? How drastic is it? What can be done to lessen the impact of a power outage on a telephone network?

- The average number of outages sufficient to cause system malfunctions per year at a typical site is approximately 15.

- 90 percent of the outages are less than 5 minutes in duration.

- 99 percent of the outages are less than 1 hour in duration.

- The total cumulative outage duration is approximately 100 minutes per year.

Availability levels of five 9s or higher require an *Uninterruptible Power Supply* (UPS) system with a minimum of one hour of battery life or a generator with an onsite service contract or a four-hour response for UPS system failures or problems. A recommended high-availability solution requires additional support to achieve five 9s overall. The HA softswitch solution must include UPS and generator backup for all distribution, core, gateway and softswitch devices. In addition, the organization should have UPS systems that have an auto-restart capability and a service contract for a four-hour response to support the UPS/generator. Given the following recommended core infrastructure, it is estimated that nonavailability (downtime due to power failure) will be two minutes per year.

The softswitch high-availability power/environment recommendations are as follows:

- UPS and generator backup

- UPS systems with an auto-restart capability

- UPS system monitoring

- Four-hour service response contract for UPS system problems

- Maintain recommended equipment operating temperatures 24/7

Overall power availability using this solution is estimated to be 99.99962 percent. This impacts overall availability, 99.99993 percent, in the same way a new module would affect a device in a serial system. The calculation used to determine overall estimated availability is then (.9999962) × (.9999993) or 99.99955 percent.[14]

# Human Error

Despite the torrent of debate over the reliability of Class 4 and 5 switches versus softswitch technologies, the chief culprit in the reliability of voice networks is human error. In the trade, it is called "procedural outage" and can mean such things as making mistakes in scheduling maintenance or planning system downtime for upgrades or other activities. Even the best-engineered networks are not immune from procedural outage. Human error presents a certain incongruity to the argument that a Class 4 or 5 switch delivers "five 9s" of reliability.

Starting with the third quarter of 1992, the FCC required the reporting of large service outages by telecommunications service providers. Reporting requirements are specified in Part 63.100 of the FCC's Rules. In an effort to monitor the reliability of the nation's telecommunications network, the *Alliance for Telecommunications Industry Solutions/Network Reliability Steering Committee* (ATIS/NRSC) has performed various statistical analyses of these FCC-reportable service outages. The results of these analyses are published in quarterly and annual reports. The following are some of the more significant findings:

- Although numerous industry initiatives have enhanced the overall quality of software and hardware used in most major network elements, there has not been a sufficient focus on procedural issues to drive improvement trends.

- On average, 33 percent of the outages reported to the FCC are attributed to procedural errors. An all-time high in this category was reached in the fourth quarter (October through December) of 1998 with 53 percent of FCC reported outages being procedural (human error) in nature.

- The distribution of outages attributed to procedural errors is 79 percent service provider and 21 percent vendor.

---

[14]Cisco Systems. "IP Telephony: The Five Nines Story." A white paper available at www.cisco.com/warp/public/cc/so/neso/vvda/iptl/5nine_wp.pdf.

- Outages attributed to procedural error are of lesser duration and impact than outages not attributed to procedural error.

- Industry-internal reports of outages attributed to procedural errors lack a consistent categorization and root-cause identification. The industry could be more effective in reducing procedural outages and enhancing network reliability if a common language and a consistent mechanism for reporting were implemented.

- Reviews of reports submitted to the FCC by service providers uncovered different formats and levels of detail, which created difficulty in understanding and categorizing some reports.

- Although some differences exist between the causes of large and small outages, they do not appear to be of such a magnitude as to influence the prioritization of solutions to procedural errors.

Two (Telcordia) generic requirements documents that have been written specifically to mitigate procedural errors. The first of these documents is GR-2914-CORE, "Human Factors Requirements for Equipment to Improve Network Reliability." This document focuses on designing equipment from a user's perspective. The objective of these proposed generic requirements is to improve the design of the maintenance user interface of network equipment by focusing on design implementations that would most likely have a significant impact on reducing procedural errors. The proposed generic requirements in this document pertain to both hardware and software interfaces between network equipment and the technicians who perform maintenance activity on the equipment. The proposed generic requirements in this document are intended to apply to new equipment manufactured as of December 1998 and may also be applied to new subcomponents or additions to existing equipment in which the hardware and software interfaces are evolving.

The spirit of GR-2914-CORE is to develop proposed generic requirements to ensure network reliability, but at the same time allow flexibility for product differentiation. The emphasis on the requirements in this document is to ensure that the design of network elements provides the necessary information for technicians to do their job and prevent confusion.

GR-454-CORE, "Generic Requirements for Supplier-Provided Documentation," presents proposed generic requirements for preparing network element documentation that will ensure that technicians have easy-to-use documentation that provides unambiguous information needed to successfully complete a task. These proposed generic requirements emphasize the importance of creating documentation from the user's perspective and

**Figure 7-8**
Wreckage of a
central office at 140
West St. across from
WTC

**Figure 7-8**
Wreckage of a
central office at 140
West St. across from
WTC

assist in promoting consistency in the content, format, and style of the documentation provided by different suppliers for similar network elements.[15]

# Conclusion

This chapter explored the assumption that Class 4 and 5 switches consistently deliver five 9s of reliability. The debate on reliability encompasses much more than the reliability of a network element (class switch versus softswitch) and must consider the network as a whole. As the events of September 11, 2001 demonstrated, the PSTN can suffer from an SPOF. Softswitch solutions, with proper engineering, can also deliver five 9s of reliability or, more precisely, availability. Given its distributed architecture, a softswitched network may prove more survivable than the PSTN as it does not suffer from an SPOF. Given Moore's Law, where computing power doubles and costs half as much every 18 months, building an HA network may prove increasingly economical. In the case of softswitch technologies, Moore's Law is very conservative in that the expansion of softswitch capabilities exceeds Moore's Law.

---

[15]ATIS/NRSC. "Procedural Outage Reduction Addressing the Human Part." A white paper available at www.atis.org/pub/nrsc/prcdrpt.pdf, pp. 3–15.

These concepts apply equally to enterprise networks, including corporate *wide area networks* (WANS). In light of the events of September 11th when so many corporate customers lost service from Verizon and other service providers, enterprise users can plan on developing a VoIP network on their corporate WAN. Deploying a softswitched VoIP solution (an IP PBX in most cases) provides an alternative for an enterprise to a Tier 1 circuit-switched service provider. It should be noted that the same guidelines apply to enterprise networks. If some 70 percent of enterprise long distance is interoffice, then it behooves an enterprise to migrate that traffic to their corporate WAN.

# Quality of Service (QoS)

A chief objection to *Voice over IP* (VoIP) and softswitched networks is the notion that their *quality of service* (QoS) is inferior to the *Public Switched Telephone Network* (PSTN). QoS covers a number of parameters, but it is mostly concerned with the user's perception of the voice quality, the call setup, and keeping the call up for the intended duration of the call (avoiding dropped calls). VoIP is time sensitive and most degradation of voice quality is the result of latency, jitter, and lost packets. It should be noted here that the context of this chapter focuses on private networks, enterprise- and service provider-maintained and not the public Internet. Voice quality in this chapter will refer to the fidelity of the reproduced speech and its intelligibility.

Great volumes have been written about QoS on *Internet Protocol* (IP) networks. This chapter will set forth measures that improve QoS on an IP network and that provide QoS, especially voice quality over an IP network, as good as that of the PSTN. Earlier in the history of the industry, a perception existed that the IP voice quality was not as good as that of circuit-switched voice. *International Telecommunication Union—Telecommunications Standardization Sector* (ITU-T) Recommendation E.800 defines QoS as "The collective effect of service performance, which determines the degree of satisfaction of a user of the service." Service providers often state that the QoS the end user experiences must equal his or her experience with the PSTN. With proper engineering, a VoIP network can deliver QoS as good or better than that of the PSTN.

# Factors Affecting QoS

The four most important network parameters for the effective transport of VoIP traffic over a softswitched network are bandwidth, delay, jitter, echo, and packet loss (see Table 8-1). Voice quality is a highly subjective thing to measure. This presents a challenge for network designers who must first focus on these issues in order to deliver the best voice quality possible. Using voice quality as a performance metric for network configurations provides appropriate mechanisms for obtaining and correlating both subjective and objective data. Perhaps the simplest test of a VoIP network is to "call your mother." That is, the simplest metric of QoS of a VoIP network is in the ear of the beholder. Just as reliability (achieving five 9s) in a network is simply a matter of good engineering, so is QoS. This chapter will explore the solutions available to service providers that will deliver the best QoS possible.

**Table 8-1**

Factors affecting
QoS

| Factor | Description |
| --- | --- |
| Delay | The time from transmission of a packet to its reception. This is measured in units of time. For VoIP transmissions that is usually a matter of milliseconds. |
| Jitter | The variation in arrival times between continuous packets transmitted from point A to point B. This is caused by packet-routing changes, congestion, and processing delays. |
| Bandwidth | Greater bandwidth delivers better voice quality. |
| Packet Loss | The percentage of packets never received at the destination. |

Just as it is necessary to closely examine every element of the network for anything that might detract from reliability when planning to deliver *High Availability* (HA), it is necessary to scrutinize the network for any element that might induce delay, jitter, packet loss, or echo. This includes first the hardware elements, such as router and media gateways, and second the routing protocols that prioritize voice packets over all other types of traffic on the IP network.

# Improving QoS in IP Routers and the Gateway

End-to-end delay is the time required for a signal generated at the talker to reach the listener. Delay is the impairment that receives the most attention in the media gateway industry. It can be corrected via functions contained in the IP network routers, the VoIP gateway, and in engineering in the IP network. The shorter the end-to-end delay, the better the perceived quality and overall user experience. The following sections discuss sources of delay.

## Sources of Delay: IP Routers

Packet delay is primarily determined by the buffering, queuing, and switching or routing delay of the IP routers. Packet capture delay is the time required to receive the entire packet before processing and forwarding it through the router. This delay is determined by the packet length, link-layer operating parameters, and transmission speed. Using short packets over high-speed trunks can easily shorten the delay. VoIP networks use packetization rates to balance connection bandwidth efficiency and packet delay.

Switching or routing delay is the time it takes a network element to forward a packet. New IP switches can significantly speed up the routing process by making routing decisions and forwarding the traffic in hardware devices instead of software. Due to the statistical multiplexing nature of IP networks and the asynchronous nature of packet arrivals, some delay in queuing is required at input and output ports of a packet switch. Overprovisioning router and link capacities can reduce this delay in queuing time.

## Sources of Delay: VoIP Gateways

Voice-signal processing at the sending and receiving ends adds to the delay and includes the time required to encode or decode the voice signal from analog or digital form into the voice-coding scheme selected for the call and vice versa. Compressing the voice signal also increases the delay. The greater the compression, the greater the delay. When bandwidth costs are not a concern, a service provider can utilize G.711, which is uncompressed voice, and this imposes a minimum of delay due to the lack of compression. Later in the chapter the parameters for G.711 and other voice-coding standards will be discussed.

On the transmit side, packetization delay is another factor that must be entered into the calculations. The packetization delay is the time it takes to fill a packet with data. The larger the packet size, the more time is required. Using shorter packet sizes can shorten this delay but will increase the overhead because more packets have to be sent, all containing similar information in the header. Balancing voice quality, packetization delay, and bandwidth utilization efficiency is very important to the service provider.[1]

How much delay is too much? Of all the factors discussed in Chapter 4, "Voice over Internet Protocol," that degrade VoIP, latency, or delay, is the greatest. Recent testing by Mier Labs offers a metric as to how much latency is acceptable or comparable to "toll quality," the voice quality offered by the PSTN. Latency less than 100 milliseconds does not affect toll-quality voice. However, latency over 120 milliseconds is discernable to most callers,

---

[1]Douskalis, Bill. *IP Telephony: The Integration of Robust VoIP Services.* New York: McGraw-Hill, 2000, pg. 230–231.

and at 150 milliseconds the voice quality is noticeably impaired, resulting in less than a toll-quality communication.

The challenge for VoIP service providers and their vendors is to get the latency of any conversation on their network to not exceed 100 milliseconds.[2] Humans are intolerant of speech delays of more than about 200 milliseconds. As mentioned earlier, ITU-T G.114 specifies that delay is not to exceed 150 milliseconds one way or 300 milliseconds round trip. The dilemma is that although elastic applications (email for example) can tolerate a fair amount of delay, they usually try to consume every bit of network capacity they can. In contrast, voice applications need only small amounts of the network, but that amount has to be available immediately (see Figure 8-1).[3]

## Other Gateway Improvements

Gateways can be engineered to minimize impairments to QoS. Those impairments are echo, end-to-end delay, buffering delay, and silence suppression. Echo is a phenomenon where a transmitted voice signal gets reflected due to an unavoidable impedance mismatch and a four-wire/two-wire conversion between the telephone handset and the communication network. Echo can disrupt the normal flow of conversation and its severity depends on the round-trip time delay. If a round-trip time delay is more

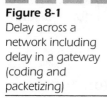

**Figure 8-1**
Delay across a network including delay in a gateway (coding and packetizing)

Sample          Encode (45 msec)    Packetize (60 msec) →

N
e
t
w
o
r
k
>100 msec

← Output        Decode (45 msec) Reconstruct (<60 msec)

[2]Mier Communications. "Lab Report—QoS Solutions." February 2001, pg. 2. www.sitaranetworks.com/solutions/pdfs/mier_report.pdf.
[3]McCullough, John and Daniel Walker. "Interested in VoIP? How to Proceed." *Business Communications Review*. April 1999, pp. 16–22.

than 30 milliseconds, the echo becomes significant, making normal conversation difficult. A gateway should use an echo canceller so that when delay reaches above 30 milliseconds, the echo canceller circuits can control the echo.

Sometimes, due to a delay in transit, some cells might arrive late. In order to ensure that no underruns occur, the buffer size should exceed the maximum predicted delay. The size of the buffer translates into delay, as each packet must progress through the buffer at the receiving gateway at the emulated circuit's line rate.

Voice communication is half-duplex, which means that one person is silent while the other speaks. A gateway can save bandwidth by halting the transmission of cells at the gateway during these silent periods. This is known as *silence suppression* or *voice activation detection* (VAD). Gateways can also offer "comfort noise generation" that approximates the buzz found on the PSTN that lets users duplicate the PSTN experience with which they are most comfortable. Some users find this silence disorienting. Many gateways offer VAD and comfort noise generators.[4]

Recent research performed by the Institute for Telecommunications Sciences in Boulder, Colorado, compared the voice quality of traffic routed through VoIP gateways with the PSTN. Researchers were fed a variety of voice samples and were asked to determine if the sample originated with the PSTN or from the VoIP gateway traffic. The result of the test was that the voice quality of the VoIP-gateway-routed traffic was "indistinguishable from the PSTN."[5] It should be noted that the IP network used in this test was a closed network and not the public Internet or another long-distance IP network. This report indicates that quality media gateways can deliver QoS on the same level as the PSTN. The challenge then shifts to ensuring the IP network can deliver similar QoS (see Figure 8-2).

## Perceptual Speech Quality Measurement (PSQM)

Another means of testing voice quality is known as PSQM. It is based on ITU-T Recommendation P.861 that specifies a model to map actual audio

---

[4]Telica. "Accelerating the Deployment of Voice over IP (VoIP) and Voice over ATM (VoATM)." A white paper posted by International Engineering Consortium, www.iec.org.
[5]Craig, Andrew. "Qualms of Quality Dog Growth of IP Telephony." *Network News*. November 11, 1999, pg. 3.

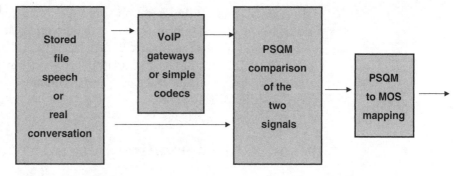

**Figure 8-2**
How Perceptual Speech Quality Measurement (PSQM) scores are determined

signals to their representations inside the head of a human. Voice quality consists of a mix of objective and subjective parts, and it varies widely among the different coding schemes and the types of network topologies used for transport. In PSQM, a measurement of processed (compressed, encoded, and so on) signals derived from a speech sample is collected, and an objective analysis is performed, comparing the original and the processed version of the speech sample. From that, an opinion is rendered as to the quality of the signal-processing functions that processed the original signal. Unlike MOS scores, PSQM scores result in an absolute number, not a relative comparison between the two signals.[6]

# Improving QoS on the Network

QoS requires the cooperation of all logical layers in the IP network—from application to physical media—and of all network elements from end to end. Clearly, optimizing QoS performance for all traffic types on an IP network presents a daunting challenge. To partially address this challenge, several *Internet Engineering Task Force* (IETF) groups have been working on standardized approaches for IP-based QoS technologies. The IETF's approaches fall into the following categories:

- Prioritization using the *Resource Reservation Setup Protocol* (RSVP) and *Differentiated Services* (DiffServ)
- Label switching using *multiprotocol label switching* (MPLS)
- Bandwidth management using the subnet bandwidth manager

[6]Douskalis, p.242–243.

To greatly simplify the objection that VoIP voice quality is not equal to the PSTN, the objection is overcome by engineering the network to diminish delay and jitter by instituting RSVP, DiffServ, and/or MPLS on the network. The International Softswitch Consortium in its reference architecture (covered in a later chapter) recommends using such tools to ensure QoS for VoIP networks.

## Resource Reservation Setup Protocol (RSVP)

A key focus in this industry is to design IP networks that will prioritize voice packets. One of the earliest initiatives is *Integrated Services* (IntServ), developed by the IETF and characterized by the reservation of network resources prior to the transmission of any data. RSVP, defined in RFC 2205, is the signaling protocol that is used to reserve bandwidth on a specific transmission path. RSVP is designed to operate with routing protocols *Open Shortest Path First* (OSPF) and the *Border Gateway Protocol* (BGP). The IntServ model is comprised of the Signaling Protocol (RSVP), an admission control routine (which determines network resource availability), a classifier (which puts packets in specific queues), and a packet scheduler (schedules packets to meet QoS requirements). The latest development is the *Resource Reservation Setup Protocol—Traffic Engineering* (RSVP-TE), a control/signaling protocol that can be used to establish a traffic-engineered path through the router network for high-priority traffic. This traffic-engineered path can operate independent of other traffic classes.

An initiative from the IEEE, 802.1p is a specification that provides a method to allow preferential queuing and access to media resources by traffic class on the basis of a "priority" value signaled in the frame. This value will provide across the subnetwork a consistent method for Ethernet, token ring, or other *media access control* (MAC) layer media types. The priority field is defined as a 3-bit value, resulting in a range of values between 0 and 7, with 0 assigned as the lowest priority and 7 indicating the highest priority. Packets may then be queued based on their relative priority values.[7]

RSVP currently offers two levels of service. The first level is guaranteed, which comes as close as possible to circuit emulation. The second level is controlled load, which is equivalent to the service that would be provided in a best-effort network under no-load conditions.

Reservation, allocation, and policing are mechanisms available in conventional packet-forwarding systems that can differentiate and appropriately handle isochronous traffic (see Table 8-2). What makes RSVP wonderfully simple in providing improved QoS for VoIP is that a network

**Table 8-2**

Reservation, alloca-
tion, and policing

| Mechanism | Description |
|---|---|
| RSVP | This provides reservation setup and control to enable the resource reservation that integrated services prescribe. Routers use RSVP to deliver QoS requests to routers along datastream paths and to maintain router and host states to provide the requested service, usually bandwidth and latency. |
| Real-Time Protocol (RTP) | This offers another way to prioritize voice traffic. Voice packets usually rely on the user datagram protocol with RTP headers. RTP treats a range of UDP ports with strict priority. |
| Committed Access Rate (CAR) | A traffic-policing mechanism, this allocates bandwidth commitments and limitations to traffic sources and destinations while specifying policies for handling traffic that exceeds the bandwidth allocation. Either the network's ingress or application flows can apply CAR thresholds. |

Source: IEEE

manager can invoke reservation, allocation, and policing of bandwidth in the RSVP protocol to improve QoS. Bandwidth is life for QoS. The better bandwidth is managed, particularly under peak load scenarios, the better the QoS, and ultimately the better the voice quality in a VoIP transmission.

RSVP works when a sender first issues a PATH message to the far end via a number of routers. The PATH message contains a *traffic specification* (Tspec) that provides details about the data packet size. Each RSVP-enabled router along the way establishes a path state that includes the previous source address of the PATH message. The receiver of the PATH message responds with a *reservation request* (RESV) that includes a *flow specification* (flowspec). The flowspec includes a Tspec and information about the type of reservation service requested, such as controlled-load service or guaranteed service.

The RESV message travels back to the sender along the same route that the PATH message took (in reverse). At each router, the requested resources are allocated, assuming that they are available and that the receiver has the authority to make the request. Finally, the RESV message reaches the sender with a confirmation that resources have been reserved.[8]

---

[7]Agarwal, Anjali. "Quality of Service (QoS) in the New Public Network Architecture." *IEEE Canadian Review*. Fall 2000.

[8]Collins, Daniel. *Carrier Grade Voice over IP*. New York: McGraw-Hill, 2000. pp. 362–363.

Guaranteed service (as opposed to controlled load; see RFC 2212) involves two elements. The first ensures that no packet loss will take place. The second ensures minimal delay. Ensuring against packet loss is a function of the token bucket depth ($b$) and the token rate ($r$) specified in the Tspec. At a given router, provided that the buffer space of value $b$ is allocated to a given flow and that a bandwidth of $r$ or greater is assigned, then there should be little to no loss. Hence, uncompressed voice will usually deliver better QoS than compressed voice.

Delay is a function of two components. The first is a fixed delay due to the processing within the individual nodes and is only a function of the path taken. The second component of delay is the queuing delay within the various nodes. Numerous mechanisms have been designed to make queuing as efficient as possible. They are described in Table 8-3. Queuing is an IP-based QoS mechanism that is available in conventional packet-forwarding systems. Queuing can differentiate and appropriately handle isochronous traffic.

Controlled load service (see RFC 2211) is a close approximation of the QoS that an application would receive if the data were being transmitted

**Table 8-3**

Queuing mechanisms

| Mechanism | Description |
|---|---|
| First-in, First-out (FIFO) | Also known as the best-effort service class, this simply forwards packets in the order of their arrival. |
| Priority Queuing (PQ) | This enables prioritization based on some defined criteria or policy. Multiple queues (high, medium, normal, and low) are filled with arriving packets according to the policies defined. *DiffServ Code Point* (DSCP) packet marking can be used to prioritize such traffic. |
| Custom Queuing (CQ) | This allocates a specific amount of a queue to each class while leaving the rest of the queue to be filled in a round-robin fashion. It essentially facilitates the prioritization of multiple classes in queuing. |
| Weighted Fair Queuing (WFQ) | This schedules interactive traffic to the front of the queue to reduce the response time and then fairly shares the remaining bandwidth among high-bandwidth flows. |
| Class-Based Weighted Fair Queuing (CBWFQ) | This combines CQ and WFQ. This strategy gives higher weight to higher-priority traffic, defined in classes using WFQ processing. |
| Low-Latency Queuing (LLQ) | This brings strict priority queuing to CBWFQ. It gives delay-sensitive data (voice) preferential treatment over other traffic. This mechanism forwards delay-sensitive packets ahead of packets in other queues. |

over a network that was lightly loaded. A high percentage of packets will be delivered successfully and the delay experienced by a high percentage of the packets will not exceed the minimum delay experienced by any successfully delivered packet.

# Differentiated Service (DiffServ)

A follow-up on the IETF initiative to IntServ was DiffServ (see RFC 2474). DiffServ sorts packets that require different network services into different classes. Packets are classified at the network ingress node according to *service level agreements* (SLAs). DiffServ is a set of technologies proposed by the IETF to allow Internet and other IP-based network service providers to offer differentiated levels of service to individual customers and their information streams. On the basis of a DSCP marker in the header of each IP packet, the network routers apply differentiated grades of service to various packet streams, forwarding them according to different *per-hop behaviors* (PHBs). The preferential *grade of service* (GoS), which can only be attempted and not guaranteed, includes a lower level of packet latency as those packets advance to the head of a packet queue, should the network suffer congestion.[9]

DiffServ makes use of the IP version 4 *Type of Service* (ToS) field and the equivalent IP version 6 Traffic Class field. The portion of the ToS/Traffic Class field that DiffServ uses is known as the DS field. The field is used in specific ways to mark a given stream as requiring a particular type of forwarding. The type of forwarding to be applied is PHB, of which DiffServ defines two types: *expedited forwarding* (EF) and *assured forwarding* (AF).

PHB is the treatment that a DiffServ router applies to a packet with a given DSCP value. A router deals with multiple flows from many sources to many destinations. Many of the flows can have packets marked with a DSCP value that indicates a certain PHB. The set of flows from one node to the next that share the same DSCP codepoint is known as an aggregate. From a DiffServ perspective, a router operates on packets that belong to specific aggregates. When a router is configured to support a given PHB, then the configuration is established in accordance with aggregates rather than to specific flows from a specific source to a specific destination.

EF (RFC 2598) is a service in which a given traffic stream is assigned a minimum departure rate from a given node, that is, one that is greater than

---

[9]Agarwal.

the arrival rate at the same node. The arrival rate must not exceed a pre-arranged maximum. This process ensures that queuing delays are removed. As queuing delays are the chief cause of end-to-end delay and are the main cause of jitter, this process ensures that delay and jitter are minimized. The objective is to provide low loss, low delay, and low latency such that the service is similar to a virtual leased line. EF can provide a service that is equivalent to a virtual leased line.

The EF PHB can be implemented in a network node in a number of ways. Such a mechanism could enable the unlimited preemption of other traffic such that EF traffic always receives access first to outgoing bandwidth. This could lead to unacceptably low performance for non-EF traffic through a token bucket limiter. Another way to implement the EF PHB would be through the use of a weighted round-robin scheduler, where the share of the output bandwidth allocated to EF traffic is equal to a configured rate.

AF (RFC 2597) is a service in which packets from a given source are forwarded with a high probability, assuming the traffic from the source does not exceed a prearranged maximum. If it does exceed that maximum, the source of the traffic runs the risk that the data will be lumped in with normal best-effort IP traffic and will be subject to the same delay and loss possibilities. In a DiffServ network, certain resources will be allocated to certain behavior aggregates, which means that a smaller share is allocated to standard best-effort traffic. Receiving best-effort service in a DiffServ network could be worse than receiving best-effort service in a non-DiffServ network. A given subscriber to a DiffServ network might want the latitude to occasionally exceed the requirements of a given traffic profile without being too harshly penalized. The AF PHB offers this possibility.

The AF PHB also enables a provider to offer different levels of forwarding assurances for packets received from a customer. The AF PHB enables packets to be marked with different AF classes and within each class to be marked with different drop-precedence values. Within a router, resources are allocated according to the different AF classes. If the resources allocated to a given class become congested, then packets must be dropped. The packets to be dropped are those that have higher drop-precedence values. The AF PHB's objective is to provide a service that ensures high-priority packets are forwarded with a greater degree of reliability than packets of a lower priority.

AF defines four classes with each class allocated a certain amount of resources (buffer space and bandwidth) within a router. Within each class, a given packet can have one of three drop rates. At a given router, if congestion exists within the resources allocated to a given AF class, the packets with the highest drop rate will be discarded first so that packets with a lower drop rate value will receive some protection. In order to function properly, the incoming traffic must not have packets with a high percentage of low drop rates. After all, the purpose is to ensure that the highest-priority packets get through in the case of congestion. That cannot happen if all the packets have the highest priority.[10]

In a DiffServ network, the AF implementation must detect and respond to long-term congestion by dropping packets. Then, it must respond to short-term congestion, which delivers a smoothed long-term congestion level. When the smoothed congestion level is below a particular threshold, no packets should be dropped. If the smoothed congestion level is between a first and second threshold level, then packets with the highest drop precedence level should be dropped. As the congestion level rises, more high drop precedence packets should be dropped until a second congestion threshold is reached. At that point, all the high drop precedence packets are dropped. If the congestion continues to rise, then packets of the medium drop precedence level should also start being dropped.

The implementation must treat all packets within a given class and precedence level equally. If 50 percent of packets in a given class and precedence value are to be dropped, then that 50 percent should be spread evenly across all packets for that class and precedence. Different AF classes are treated independently and are given independent resources. When packets are dropped, they are dropped for a given class and drop precedence level. Packets of one class and precedence level might possibly experience a 50 percent drop rate while the packets of a different class with the same precedence level are not dropped at all. Regardless of the amount of packets that need to be dropped, a DiffServ node must not reorder AF packets within a given AF class, regardless of their precedence level.[11]

## MPLS-Enabled IP Networks

MPLS has emerged as the preferred technology for providing QoS, traffic engineering, and VPN capabilities on the Internet. MPLS contains forwarding information for IP packets that is separate from the content of the IP header such that a single forwarding paradigm (label swapping) operates in conjunction with multiple routing paradigms. The basic operation of

[10]Collins, pg. 384.

[11]Ibid., pp. 386–387.

MPLS is to establish *label-switched paths* (LSPs) into which certain types of traffic are directed. MPLS provides the flexibility of being able to form *Forwarding Equivalence Classes* (FECs) and the capability to create a forwarding hierarchy via label stacking. All these techniques facilitate the operation of QoS, traffic engineering, and VPNs. MPLS is similar to Diff-Serve in that it marks traffic at the entrance to the network. The function of the marking is to determine the next router in the path from the source to the destination.

MPLS involves the attachment of a short label to a packet in front of the IP header. This procedure is effectively similar to inserting a new layer between the IP layer and the underlying link layer of the *Open Systems Interconnection* (OSI) model. The label contains all the information that a router needs to forward a packet. The value of a label can be used to look up the next hop in the path for forwarding to the next router. The difference between this routing and standard IP routing is that the match is exact. This enables faster routing decisions in routers.[12]

An MPLS-enabled network, on the other hand, is able to provide low latency and guaranteed traffic paths for voice. Using MPLS, voice traffic can be allocated to an FEC that provides the differentiated service appropriate for this traffic type. Significant work has been done recently to extend MPLS as the common control plane for optical networks.[13]

QoS in softswitched networks is corrected with mechanisms similar to those in TDM networks. By engineering out deficiencies in the components (media gateways) and improving the network (DiffServ and MPLS), QoS can be brought up to the standards of the PSTN. Although not as quantifiable as a *Mean Opinion Score* (MOS) score on a media gateway, significant progress has been made in recent years in engineering closed IP networks to deliver PSTN quality voice. Figure 8-3 illustrates the progression in improving QoS on those networks.

MPLS is not primarily a QoS solution. It is a new switching architecture. Standard IP switching requires every router to analyze the IP header and make a determination of the next hop based on the content of that header. The primary driver in determining the next hop is the destination address in the IP header. A comparison of the destination address with entries in a routing table and the longest match between the destination address and

---

[12]Ibid., pg. 364.

[13]Integral Access. "The Evolution Toward Multiservice IP/MPLS Networks." A white paper located at www.integralaccess.com, 2001. pp. 4–5.

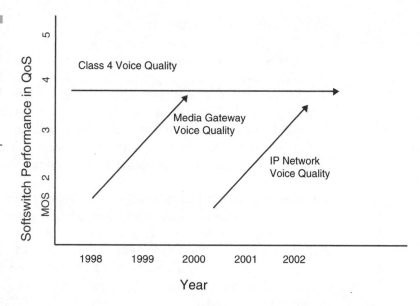

**Figure 8-3**
Voice quality for
media gateways and
IP networks versus
Class 4 (PSTN). Media
gateways now score
4.0 or better in MOS
testing.

the addresses in the routing table determines the next hop. The approach with MPLS is to attach a label to the packet. The content of the table is specified according to an FEC, which is determined at the point of ingress to the network. The packet and label are passed to the next node where the label is examined and the FEC is determined. This label is then used as a simple lookup in a table that specifies the next hop and a new label to use. The new label is attached and the packet is forwarded.

The major difference between label switching and standard routing based on IP is that the FEC is determined at the point of ingress to the network where information might be available that cannot be indicated in the IP header. The FEC can be chosen based on a combination of the destination address, QoS requirements, the ingress router, or a variety of other criteria. The FEC can indicate such information and routing decisions in the network, automatically taking that information into account. A given FEC can force a packet to take a particular route through the network without having to cram a list of specific routers into the IP header. This is important for ensuring QoS when the bandwidth that is available on a given path has a direct impact on the perceived quality.[14]

---

[14]Collins, p. 399.

# MPLS Architecture

MPLS assigns an FEC value to a packet upon its ingress to the network. That FEC value is then mapped to a particular label value and the packet is forwarded with the label. At the next router, the label is evaluated and a corresponding FEC is determined. A lookup is then performed to determine the next hop and a new label to apply. The new label is attached and the packet is forwarded to the next node. This process indicates that the value of the label can change as the packet moves through the network.

**Label-Switching Routers (LSRs)**   The relationship between the FEC and the label value is a local affair between two adjacent *label-switching routers* (LSRs). If a given router is upstream from the point of view of data flow, then it must have an understanding with the next router downstream as to the binding between a particular label value and a particular FEC.

An LSR's actions depend on the value of the label. The LSR's action is specified by the *Next Hop-Level Forwarding Entry* (NHLFE), which indicates the next hop, the operation to perform on the label stack, and the encoding to be used for the stack on the outgoing link. The operation to perform on the stack might mean that the LSR should replace the label at the top of the stack with a new label. The operation might require the LSR to pop the label stack or to replace the top label with a new label, rather than add one or more additional labels on top of the first label.

The next hop for a given labeled packet might be the same LSR. In such a case, the LSR pops the top-level label of the stack and forwards the packet to itself. At this point, the packet might still have a label to be examined, or it might be a native IP packet without a label (in which case, the packet is forwarded according to standard IP routing).

A given label might possibly map to more than one NHLFE. This might occur when load sharing takes place across multiple paths. Here the LSR chooses one NHLFE according to internal procedures. When a router knows it is the next-to-last LSR in a given path, it removes labels and passes the packet to the final LSR sans the label. This is done to streamline the work of the last router. If the next-to-last LSR passes a labeled packet to the final LSR, then the final LSR must examine the label, determine that the next hop is itself, pop the stack, and forward the packet to itself. The LSR must then reexamine the packet to determine what to do with it. If the packet arrives without a label, then the final LSR has one less step to execute. How a particular LSR determines that it is the next-to-last LSR for a given path is a function of label distribution and the distribution protocol used.[15]

**Label-Switched Paths (LSPs)**    MPLS networks are subsets of a larger IP network. This means there will be points of ingress and egress to the MPLS networks from the larger IP network. An LSR that is a point of ingress to the MPLS network will be responsible for choosing the FEC that should be applied to a given packet. As label distribution works in a downstream-to-upstream direction, an LSR that is a point of egress is responsible for determining a label/FEC binding and passing that information upstream. An LSR will act as an egress LSR with respect to a particular FEC if the FEC refers to the LSR itself, if the next hop for the FEC is outside the label-switching network, or if the next hop means traversing a boundary. An LSP is the path to a certain FEC from a point of ingress to the egress LSR. The primary function of *label-distribution protocols* (LDPs) is the establishment and maintenance of these LSPs.

**Label-Distribution Protocols (LDPs)**    In the MPLS architecture, the downstream LSR decides on the particular binding. The downstream LSR then communicates the binding to the upstream LSR, which means that an LDP must be established between the two to support such communication. Label distribution is performed in two ways. First, a downstream on demand exists where a given LSR can request a particular label/FEC binding from a downstream LSR. Second, an unsolicited downstream exists, where a given LSR distributes label/FEC bindings to other upstream LSRs without having been explicitly requested to do so.

## MPLS Traffic Engineering

Performance objectives for VoIP networks are either traffic oriented or performance oriented. Traffic-oriented objectives deal with QoS and aim to decrease the impacts of delay, jitter, and packet loss. Performance-oriented objectives seek to make optimum usage of network resources, specifically network bandwidth. Congestion avoidance is a major objective related to both network resource objectives and QoS objectives. As regards resource

---

[15]Ibid.

objectives, it is imperative to avoid having one part of a network being congested while another part of the network is underutilized, where the underutilized part could carry traffic from the congested part of the network. From a QoS perspective, it is necessary to allocate traffic streams where resources are available to ensure that those streams do not experience congestion with resultant packet loss and delay.

Congestion occurs in two ways. First, it results from a lack of adequate resources on the network to handle the offered load. Second, it occurs from the steering of traffic towards resources that are already loaded as other resources remain underutilized. The expansion of flow control can correct the first situation. Good traffic engineering can overcome limitations of steering traffic to avoid congestion. Current IP routing and resource allocation is not well equipped to deal with traffic engineering.

MPLS offers the concept of the traffic trunk, which is a set of flows that share specific attributes. These attributes include the ingress and egress LSRs, the FEC, and other characteristics such as average rate, peak rate, and priority and policing attributes. A traffic trunk can be routed over a given LSP. The LSP that a traffic trunk would use can be specified. This allows certain traffic to be steered away from the shortest path, which is likely to be congested before other paths. The LSP that a given traffic trunk will use can be changed. This enables the network to adapt to changing load conditions either via administrative intervention or through automated processes within the network.

Traffic engineering on an MPLS network has the main elements of the mapping of packets to an FEC, mapping FECs to traffic trunks, and mapping traffic trunks onto the physical network topology through LSPs. The assignment of individual packets to a given FEC and how those FECs are further assigned to traffic trunks are functions specified at the ingress to the network. These decisions can be made according to various criteria, provided they are understood by both the MPLS network provider and the source of packets (the customer or other network provider).

A third mapping that must take place revolves around providing the quality that is needed for a given type of traffic. This mapping involves constraint-based routing, where traffic is matched with network resources according to the characteristics of the traffic and of available resources. That is, one characteristic of traffic is the bandwidth requirements, and one characteristic of a path is the maximum bandwidth that it offers.

To date, MPLS is considered the best means of engineering an IP network to handle voice traffic to deliver the best possible QoS. As this technology becomes more widely deployed in IP networks, VoIP will be delivered with a quality at least equal to the PSTN.[16]

# Measuring Voice Quality

How does one measure the difference in quality between a VoIP network and the PSTN? As the VoIP industry matures, new means of measuring voice quality are arriving on the market. Currently, two tests award some semblance of a score for voice quality. The first is a holdover from the circuit-switched voice industry known as *mean opinion score* (MOS). The other has emerged with the rise in popularity of VoIP and is known as PSQM.

## Mean Opinion Score (MOS)

Can QoS be measured scientifically? The telephone industry employs a subjective rating system known as the MOS to measure the quality of the telephone connections. The measurement techniques are defined in ITU-T P.800 and are based on the opinions of many testing volunteers who listen to a sample of voice traffic and rate the quality of that transmission. The volunteers listen to a variety of voice samples and are asked to consider factors such as loss, circuit noise, side tone, talker echo, distortion, delay, and other transmission problems. The volunteers then rate the voice samples from 1 to 5, with 5 being excellent and 1 being bad. The voice samples are then awarded an MOS. An MOS of 4 is considered toll quality. Table 8-4 shows the MOS scores of speech codecs in the PSTN. Table 8-5 illustrates how MOSs for some gateways now meet or exceed the 4.0 score assigned to the PSTN.

---

[16]Ibid.

**Table 8-4**

MOS scores of speech codecs in the PSTN

| Standard | Data Rate (Kbps) | Delay (ms) | MOS |
|---|---|---|---|
| G.711 | 64 | 0.125 | 4.8 |
| G.721<br>G.723<br>G.726 | 16,24,32,40 | 0.125 | 4.2 |
| G.728 | 16 | 2.5 | 4.2 |
| G.729 | 8 | 10 | 4.2 |
| G.723.1 | 5.3, 6.3 | 30 | 3.5, 3.98 |

**Table 8-5**

Comparison of MOS scores for VoIP gateways and Class 4 switches

| Vendor and product | MOS |
|---|---|
| Nortel DMS-250 (Class 4) | 4.0 |
| Lucent 4ESS (Class 4) | 4.0 |
| Convergent ICS2000 (Softswitch) | 4.0 |
| SONUS GSX9000 (Softswitch) | 4.0 |
| Nuera Nu-Tandem (Softswitch) | 4.83 |

# Conclusion

Adequate QoS requires optimizing queuing strategies, call-admission controls, congestion-avoidance mechanisms, and traffic-shaping and policing technologies. Given an IP network system's distributed nature, administrators must often perform these often highly interrelated optimizations simultaneously. Different QoS technologies, each addressing different aspects of end-to-end performance and each implemented in particular ways by equipment vendors, must be evaluated to achieve an adequate QoS for VoIP deployments.[17] A properly engineered and managed IP network can deliver voice quality equal to or better than the PSTN.

---

[17]Sheik, Fayez, Stan McClellan, Manpreet Sing, and Sannedi Chakravarthy. End-to-End Testing of IP QoS. A white paper from the IEEE, available online at www.computer.org/computer/co2002/r5080abs.htm

# SS7 and Softswitch

The 5ESS Class 5 switch from Lucent Technologies reportedly offers some 3,500 features. The features are a function of *Signaling System 7* (SS7) signaling (refer back to Chapter 2, "The Public Switched Telephone Network (PSTN)," for an SS7 tutorial). If the SS7 signaling network doesn't work, the 3,500 features will not be available to the subscriber. Almost every nation in the world has its own version of SS7 (or C7 outside the United States). Any international call must intermediate between two or more versions of SS7 in order for the call to be completed. In short, given the level of complexity of multiple SS7 versions, it's a miracle any international call is completed. The rise of VoIP with its own signaling protocols (H.323 and SIP) presents another challenge when interconnecting *Public Switched Telephone Network* (PSTN) and *Voice over IP* (VoIP) calls. The grand challenge for *Internet telephony service providers* (ITSPs) is the intermediation of *Internet Protocol* (IP) networks and the PSTN. Such intermediation must not only perform call setup and teardown, but it must also transmit the signaling necessary for call features to be used. This chapter will describe how SS7 is transported over IP networks to facilitate signaling between the PSTN and VoIP networks.

A VoIP network carries voice traffic cheaper than a switched-circuit telephone network because IP telephony networks make better use of available bandwidth. In a public switched telephone network, for example, a dedicated 64 Kbps end-to-end circuit is allocated for each call. In a VoIP network, digitized voice data is highly compressed and carried in packets over IP networks. Using the same bandwidth, a VoIP network can carry many times the number of voice calls as a switched circuit network with better voice quality. The savings realized in using VoIP networks are often passed onto users in the form of lower costs.

In 2001, U.S. telephone companies are expected to offload between 15 and 20 percent of overseas voice calls to VoIP network operators, such as iBasis, and the percentage is rising each year as the VoIP network infrastructure is rolled out. Other countries, such as China, carry an even higher percentage of voice traffic over domestic and international VoIP networks.

Metcalfe's Law (named for Bob Metcalfe, inventor of Ethernet) states that the value of a network rises in proportion to the power of all the machines attached to it. In a telephone network, the number of telephones determines the value attached to the network. A private, enterprise VoIP network may be limited to the number of telephones or telephone-equivalent endpoints (that is, the telephony agent on a PC). The value of that network increases drastically with the addition of a VoIP gateway that will connect all the telephones on the enterprise network with the PSTN. By being able to access every telephone in the world (that is, all those connected to the PSTN), the value of the network grows exponentially in terms of the number of other telephones that can be reached from any telephone on that network. For this to happen, it is critical that the signaling systems of the private, VoIP enterprise telephone network interoperate with the PSTN. To do this, it is necessary for the VoIP network to transport SS7. Figure 9-1 provides some PSTN signaling examples and Figure 9-2 shows an example of signaling on a VoIP network.

Fortunately, much work has been done in solving this dilemma. Every month, billions of minutes of long-distance traffic travels between IP networks and various PSTNs worldwide. If a VoIP service is to be considered carrier grade, it must support SS7. This chapter will outline the measures used to bridge IP networks and the PSTN.

**Figure 9-1**
*Signaling in the PSTN: channel-associated and common channel signaling*

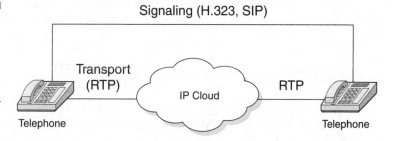

**Figure 9-2**
*Signaling on a VoIP network: common channel signaling with H.323 or SIP*

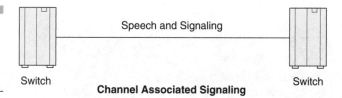

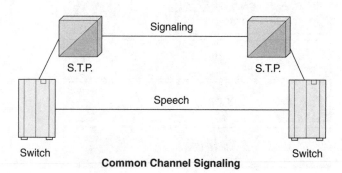

# Signaling in the PSTN (SS7 or C7)

In order for a VoIP network to interoperate with the PSTN, it must interoperate with the SS7 network. At some point in the future, all voice conversations will be via VoIP using a VoIP-specific signaling protocol such as the *Session Initiation Protocol* (SIP). However, in the present, a VoIP network must support SS7 if calls across that network originate or terminate in the PSTN. SS7 is not a protocol in and of itself, but rather a collection of subprotocols that operate together to deliver the SS7 functionality. The SS7 protocol stack, as detailed in Figure 9-3, is somewhat similar to the *Open Systems Interconnection* (OSI) model of IP networks. By understanding the SS7 protocols, it is possible to understand which SS7 elements interoperate with which VoIP signaling elements (SS7 is described in detail in Chapter 2). A review of the key elements of SS7 is provided here. The following paragraphs describe the subprotocols of SS7, as illustrated in Figure 9-3.

## Message Transfer Part (MTP)

The *Message Transfer Part* (MTP) is divided into three levels. The lowest level, MTP Level 1, is equivalent to the OSI physical layer. MTP Level 1 defines the physical, electrical, and functional characteristics of the digital signaling link. The physical interfaces defined include E-1 (2048 Kbps; 32

**Figure 9-3**
SS7 protocol stack

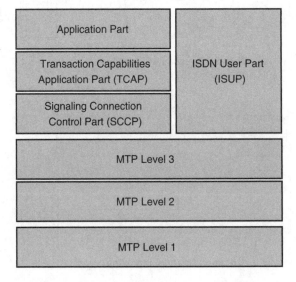

64-Kbps channels), DS-1 (1544 Kbps; 24 64-Kbps channels), V.35 (64 Kbps), DS-0 (64 Kbps), and DS-0A (56 Kbps). MTP Level 2 ensures accurate end-to-end transmission of a message across a signaling link. Level 2 implements flow control, message sequence validation, and error checking. When an error occurs on a signaling link, the message (or set of messages) is retransmitted. MTP Level 2 is equivalent to the OSI data link layer. MTP Level 3 provides message routing between signaling points in the SS7 network. MTP Level 3 reroutes traffic away from failed links and signaling points, and it controls traffic when congestion occurs. MTP Level 3 is equivalent to the OSI network layer.

## ISDN User Part (ISUP)

The *ISDN User Part* (ISUP) defines the protocol used to set up, manage, and release trunk circuits that carry voice and data between terminating line exchanges (between a calling party and a called party). ISUP is used for both *Integrated Services Digital Network* (ISDN) and non-ISDN calls. However, calls that originate and terminate at the same switch do not use ISUP signaling.

## Signaling Connection Control Part (SCCP)

*Signaling Connection Control Part* (SCCP) provides connectionless and connection-oriented network services and *global title translation* (GTT) capabilities above MTP Level 3. A global title is an address (a dialed 800 number, a calling card number, or a mobile subscriber identification number), which is translated by SCCP into a destination point code and subsystem number. A subsystem number uniquely identifies an application at the destination signaling point. SCCP is used as the transport layer for services based on the *Transaction Capabilities Applications Part* (TCAP).

## Transaction Capabilities Applications Part (TCAP)

TCAP supports the exchange of noncircuit-related data between applications across the SS7 network using the SCCP connectionless service. Queries and responses sent between *service switching points* (SSPs) and *service control points* (SCPs) (refer to Chapter 2 for more information) are carried in TCAP messages. For example, an SSP sends a TCAP query to determine the routing number associated with a dialed 800/888 number and to check the *personal identification number* (PIN) of a calling card user. In mobile networks (IS-41 and GSM), TCAP carries *Mobile Application Part* (MAP) messages sent between mobile switches and databases to support user authentication, equipment identification, and roaming.[1]

# Interworking SS7 and VoIP Networks

If SS7 is so critical to a phone call, how can a VoIP network interoperate with SS7? Very simply, a mechanism known as *Signaling Transport* (SIG-TRAN) facilitates the transport of the SS7 subprotocols over an IP network. The following section explains how each of the subprotocols of SS7 is transported over an IP network. Figure 9-4 illustrates from a high level the interaction between IP and PSTN networks for the completion of a simple phone call. Figure 9-5 details the SIGTRAN protocol stack.

**Figure 9-4**
IP network to PSTN
with SS7 signaling

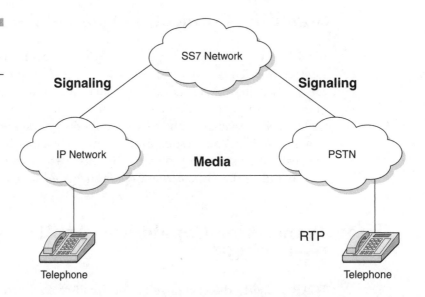

---

[1]Performance Technologies. "SS7 Tutorial." A white paper at www.pt.com/tutorials/ss7.

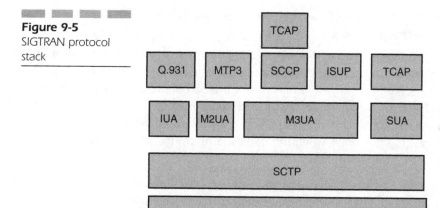

**Figure 9-5**
SIGTRAN protocol stack

# Signaling in VoIP Networks

VoIP networks carry SS7 over IP using protocols defined by the SIGTRAN working group of the *Internet Engineering Task Force* (IETF), the international organization responsible for recommending Internet standards. The SIGTRAN protocols support the stringent requirements for SS7 and C7 (the non-U.S. version of SS7) signaling, as defined by the *International Telecommunications Union—Telecommunications Standardization Sector* (ITU-T).

In IP telephony networks, signaling information is exchanged between the following functional elements (described in greater detail in Chapter 3, "Softswitch Architecture or 'It's the Architecture, Stupid!'"):

■ **Media gateway**  A media gateway terminates voice calls on interswitch trunks from the PSTN, compresses and packetizes the voice data, and delivers compressed voice packets to the IP network. For voice calls originating in an IP network, the media gateway performs these functions in reverse order. For ISDN calls from the PSTN, Q.931 signaling information is transported from the media gateway to the *media gateway controller* (MGC) for call processing.

■ **Media gateway controller (MGC, also known as gatekeeper or softswitch)**  An MGC handles the registration and management of resources at the media gateway(s). An MGC exchanges ISUP messages with central office switches via a signaling gateway. Because vendors of

MGCs often use off-the-shelf computer platforms, an MGC is sometimes called a softswitch.

- **Signaling gateway** A signaling gateway provides the transparent interworking of signaling between switched circuit and IP networks. The signaling gateway must terminate SS7 signaling or translate and relay messages over an IP network to an MGC or another signaling gateway. Because of its critical role in integrated voice networks, signaling gateways are often deployed in groups of two or more to ensure high availability.

A media gateway, signaling gateway, or MGC (softswitch) may be separate physical devices or be integrated in any combination.[2]

# Signaling Transport (SIGTRAN)

SIGTRAN architecture is defined in RFC 2719, "Framework Architecture for Signaling Transport." The architecture uses the signaling components shown in Figure 9-6. Signaling over standard IP uses a common transport protocol that ensures reliable signaling delivery and an adaptation layer

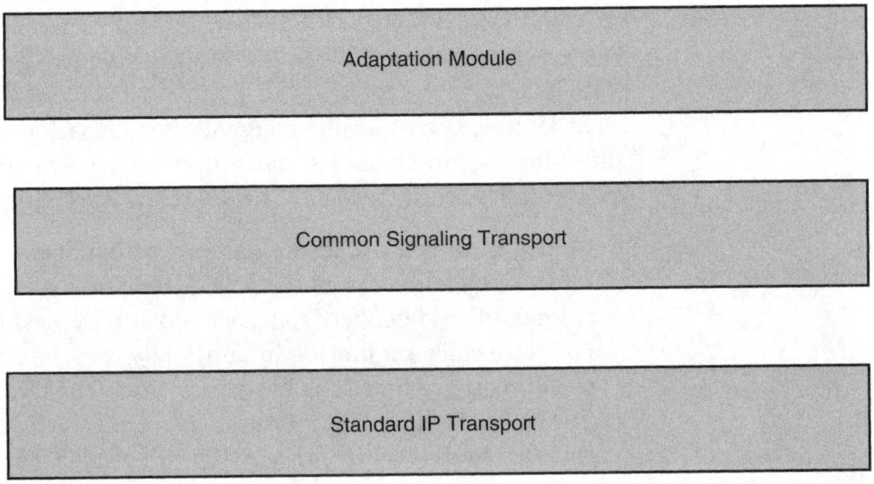

**Figure 9-6**
SIGTRAN
components
(Source: Collins)

Adaptation Module

Common Signaling Transport

Standard IP Transport

---

[2]IETF. Request for Comments 2719: "Framework Architecture for Signaling Transport." October 1999, pg. 17.

supporting specific primitives as required by a particular signaling application. That is, the common signaling transport makes sure that messages are delivered error free and in sequence, despite the inconsistencies of an IP network. The adaptation layer provides an interface to the SS7 application so that those applications do not realize that the underlying transport is IP as opposed to the SS7 network.[3]

# SIGTRAN Protocols

The SIGTRAN protocols specify the means by which SS7 messages can be reliably transported over IP networks. The architecture identifies two components: a common transport protocol for the SS7 protocol layer being carried and an adaptation module to emulate lower layers of the protocol. For example, if the native protocol is *Message Transport Layer* (MTP) Level 3, the SIGTRAN protocols provide the equivalent functionality of MTP Level 2. If the native protocol is ISUP or SCCP, the SIGTRAN protocols provide the same functionality as MTP Levels 2 and 3. If the native protocol is TCAP, the SIGTRAN protocols provide the functionality of SCCP (connectionless classes) and MTP Levels 2 and 3.

The SIGTRAN protocols provide all the functionality needed to support SS7 signaling over IP networks, including

- Flow control
- In-sequence delivery of signaling messages within a single control stream
- Identification of the originating and terminating signaling points
- Identification of voice circuits
- Error detection, retransmission, and other error-correcting procedures
- Recovery from outages of components in the transit path
- Controls to avoid congestion on the Internet
- Detection of the status of peer entities (in service, out of service, and so on)
- Support for security mechanisms to protect the integrity of the signaling information
- Extensions to support security and future requirements

---

[3]Collins, Daniel. *Carrier Grade Voice Over IP*. New York: McGraw-Hill, 2000. pg. 321.

Restrictions imposed by narrow-band SS7 networks, such as the need to segment and reassemble messages greater than 272 bytes, are not applicable to IP networks and are therefore not supported by the SIGTRAN protocols.[4]

# Stream Control Transmission Protocol (SCTP)

To reliably transport SS7 messages over IP networks, the IETF SIGTRAN working group devised the *Stream Control Transmission Protocol* (SCTP). SCTP enables the reliable transfer of signaling messages between signaling endpoints in an IP network.

**Associations**   To establish an association between SCTP endpoints, one endpoint provides the other endpoint with a list of its transport addresses (multiple IP addresses in combination with an SCTP port). These transport addresses identify the addresses that will send and receive SCTP packets.

IP signaling traffic is usually composed of many independent message sequences between many different signaling endpoints. SCTP enables signaling messages to be independently ordered within multiple streams (unidirectional logical channels established from one SCTP endpoint to another) to ensure the in-sequence delivery between associated endpoints. By transferring independent message sequences in separate SCTP streams, it is less likely that the retransmission of a lost message will affect the timely delivery of other messages in unrelated sequences (called *head-of-line blocking*).

To meet SS7 signaling reliability and performance requirements for carrier-grade networks, VoIP network operators ensure that no *single point of failure* (SPOF) exists in the end-to-end network architecture between an SS7 node and an MGC. To achieve carrier-grade reliability in IP networks, links in a linkset are typically distributed amongst multiple signaling gateways, MGCs are distributed over multiple *central processing unit* (CPU) hosts, and redundant IP network paths are provisioned to ensure the survivability of SCTP associations between SCTP endpoints.

**Packets and Chunks**   SCTP sits on top of IP in the SIGTRAN protocol stack. When SCTP wants to send a piece of information to the remote end,

---

[4]IETF, pg. 17.

it sends what is known as an SCTP packet to IP, and IP routes the packet to the destination. The SCTP packet is comprised of a common header and a number of chunks. The common header includes the source and destination port numbers that, when combined with the source and destination IP addresses, uniquely identify the endpoints. The header also includes a verification tag that is used to validate the sender of the packet. The common header also includes an Adler-32 checksum, which is a particular calculation based on the values of the octets in the packet. This checksum is used to ensure that the packet has been received without corruption and provides another level of protection above the IP header checksum.

A number of chunks enable the common header, and each chunk is comprised of a chunk header plus some chunk-specific content. This content can be either SCTP control information or SCTP user information.

**Streams** A stream is a one-way logical channel between SCTP endpoints. A stream is a sequence of user messages between two SCTP users. When an association is established between endpoints, part of the establishment of the association involves specifying how many streams the association must support. If we think of a given association as a one-way highway between endpoints, then the individual streams are analogous to the individual traffic lanes on that highway (see Figure 9-7).[5]

**Figure 9-7**
*Streams allocated according to signaling links of Circuit Identification Code (CIC) values (Source: Collins)*

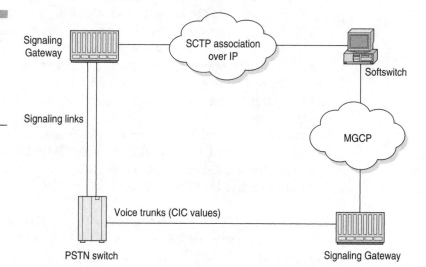

---

[5]Collins, pp. 326–327.

**SCTP Robustness** Robustness is a key characteristic of any carrier-grade network. It means that the network should implement procedures through which failures or undesired occurrences are minimized. Robustness also includes the capability to handle a certain amount of failure on the network without a significant reduction in quality. Furthermore, the network should provide a graceful degradation in the event of failures or overload. To do this, SCTP implements congestion-control mechanisms to ensure that one endpoint does not flood another with messages. It also utilizes path *maximum transmission unit* (MTU) discovery so that messages are not sent if they are too long to be handled by the intervening transport network.[6]

## Transporting MTP2 over IP: M2UA

What makes SIGTRAN so unique is the mechanism to transport SS7 protocols over IP. By accomodating the existing SS7 protocols, SIGTRAN enables SS7 signaling to operate over an IP network. By mating SS7 protocols with IP transport, SIGTRAN enables interworking of IP with TDM telephone networks.

**MTP2 User Adaptation Layer (M2UA)** The *MTP2 User Adaptation Layer* (M2UA) is a protocol defined by the SIGTRAN Working Group for transporting SS7 MTP Level 2 user signaling messages over IP using SCTP. The M2UA protocol layer provides the equivalent set of services to its users as MTP Level 2 provides to MTP Level 3.

M2UA is used between the signaling gateway and MGC in VoIP networks. The signaling gateway receives SS7 messages over an MTP Level 1 and Level 2 interface from a signaling end point (SCP or SSP) or a *signal transfer point* (STP) in the PSTN. The signaling gateway terminates the SS7 link at MTP Level 2 and transports MTP Level 3 and above to an MGC or other IP endpoint using M2UA over SCTP/IP. The signaling gateway maintains the availability state of all MGCs to manage signaling traffic flows across active SCTP associations.

**Transporting MTP3 over IP: M3UA** The *MTP3 User Adaptation Layer* (M3UA) is a protocol defined by the IETF SIGTRAN Working Group for transporting MTP Level 3 user part signaling messages (ISUP, *Telephone User Part* [TUP], and SCCP) over IP using SCTP. TCAP messages, as SCCP

---

user protocols, may be carried by SCCP using M3UA or by a different SIG-TRAN protocol called *SCCP User Adaptation Layer* (SUA), as described later.

M3UA is used between a signaling gateway and an MGC or IP telephony database. The signaling gateway receives SS7 signaling using MTP as transport over a standard SS7 link. The signaling gateway terminates MTP-2 and MTP-3 and delivers ISUP, TUP, SCCP and/or any other MTP-3 user messages, as well as certain MTP network management events over SCTP associations to MGCs or IP telephony databases.

The ISUP and/or SCCP layer at an IP signaling point is unaware that the expected MTP-3 services are not provided locally, but rather by the remote signaling gateway. Similarly, the MTP-3 layer at a signaling gateway may be unaware that its local users are actually remote parts over M3UA. Conceptually, M3UA extends access to MTP-3 services at the signaling gateway to remote IP endpoints. If an IP endpoint is connected to more than one signaling gateway, the M3UA layer at the IP endpoint maintains the status of configured SS7 destinations and routes messages according to the availability and congestion status of the routes to these destinations via each signaling gateway.

At the signaling gateway, the M3UA layer provides interworking with MTP-3 management functions to support the seamless operation of signaling between the SS7 and IP networks. For example, the signaling gateway indicates to remote MTP-3 users at IP endpoints when an SS7 signaling point is reachable or unreachable, or when SS7 network congestion or restrictions occur. The M3UA layer at an IP endpoint keeps the state of the routes to remote SS7 destinations and may request the state of remote SS7 destinations from the M3UA layer at the signaling gateway. The M3UA layer at an IP endpoint may also indicate to the signaling gateway that M3UA at an IP endpoint is congested (see Figure 9-8).

**Transporting SCCP over IP: SUA**   SUA is a protocol defined by the IETF SIGTRAN Working Group for transporting SS7 SCCP user part signaling messages (TCAP and RANAP) over IP using the SCTP. SUA is used between a signaling gateway and an IP signaling endpoint and between IP signaling endpoints. SUA supports both SCCP unordered and in-sequence connectionless services, bidirectional connection-oriented services with or without flow control, and the detection of message loss and out-of-sequence errors.

For a connectionless transport, SCCP and SUA interface at the signaling gateway. From the perspective of an SS7 signaling point, the SCCP user is located at the signaling gateway. SS7 messages are routed to the signaling gateway based on the point code and SCCP subsystem number.

**Figure 9-8**
M3UA usage in a
signaling gateway to
softswitch
applications

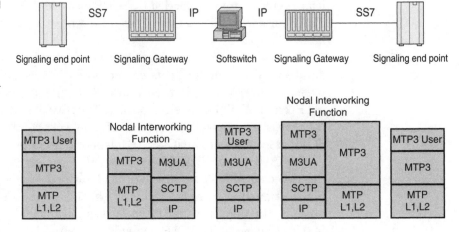

**Figure 9-8**
M3UA usage in a
signaling gateway to
softswitch
applications

The signaling gateway then routes SCCP messages to the remote IP endpoint. If redundant IP endpoints exist, the signaling gateway(s) can load share among active IP endpoints using a round-robin approach.

Note that the load sharing of TCAP messages occurs only for the first message in a TCAP dialog; subsequent TCAP messages in the same dialog are always sent to the IP endpoint selected for the first message, unless endpoints share state information and the signaling gateway is aware of the message allocation policy of the IP endpoints.

The signaling gateway may also perform *Global Title Translation* (GTT) to determine the destination of an SCCP message. The signaling gateway routes on the global title, that is, the digits present in the incoming message, such as the called party number or the mobile subscriber identification number.

For connection-oriented transport, SCCP and SUA interface at the signaling gateway to associate the two connection sections needed for a connection-oriented data transfer between an SS7 signaling end point and an IP endpoint. Messages are routed by the signaling gateway to SS7 signaling points based on the destination point code (in the MTP-3 address field) and to IP endpoints based on the IP address (in the SCTP header).

SUA can also be used to transport SCCP user information between IP endpoints directly rather than via the signaling gateway. The signaling gateway is needed only to enable interoperability with SS7 signaling in the switched circuit network.

If an IP resident application is connected to multiple signaling gateways, multiple routes may exist to a destination in the SS7 network. In this case, the IP endpoint must monitor the status of remote signaling gateways before initiating a message transfer.[7]

M3UA over SCTP can be used in a VoIP network to provide interworking with the external SS7 network for ISUP. The chief components are the *application server* (AS), *application server process* (ASP), routing key, and network appearance (see Table 9-1).

**Transporting TCAP over IP: SS7 TCAP-User Adaptation Layer (TUA)** The *TCAP-User Adaptation Layer* (TUA) is an SS7 signaling user adaptation layer for providing TCAP-user signaling over SCTP.

# ISDN Q.921-User Adaptation Layer (IUA)

In ISDN, user signaling such as Q.931 is carried by the Q.921 data link layer. The equivalent SIGTRAN specification in an IP network is IUA.

**Table 9-1**

M3UA operational components

| Component | Description |
|---|---|
| Application server (AS) | A logical entity handling signaling for a particular scope. An AS can be a logical entity within a call agent that handles ISUP signaling for a particular SS7 DPC/OPC/CIC range. |
| Application server process (ASP) | A process instance of an application server acting as an active or standby process. An ASP could be that process within a softswitch that is handling ISUP signaling. The ASP has an SCTP endpoint and it can be spread across multiple IP addresses. In a robust network, at least one ASP should be active and at least one standby ASP should exist for a given application. |
| Routing key | A set of SS7 parameters that identifies the signaling for a given AS. If a given AS is to handle ISUP signaling for a particular combination of an OC/DPC/CIC range, then that range is the routing key. Within the AS, at least one ASP will actively handle the signaling. |
| Network appearance | A mechanism for separating signaling traffic between an SG and an ASP where all traffic uses the same underlying SCTP association. |

---

[7]Performance Technologies. "Signaling in Switched Circuit and VoIP Networks." www.pt.com/tutorials/iptelephony/tutorial_voip_signaling.html.

Q.921 messages can be passed from the ISDN to the IP network with identical Q.931 implementations in each network, neither of them recognizing any difference in the underlying transport.[8]

## Signaling Network Architecture

To ensure carrier-grade service, it is necessary to ensure no SPOF exists on the network between the signaling gateway and the ASP. Signaling gateways are deployed in pairs similar to STPs in an SS7 network to provide redundancy. ASPs are software hosted on a variety of servers to provide survivability. Chapter 7, "Softswitch Is Just as Reliable as Class 4/5 Switches," covered engineering networks for high availability.

Each ASP requires a point code. ASPs usually have the same point code as their associated signaling gateway. This way, the ASP and signaling gateway appear to the SS7 networks as a single signaling endpoint (see Figure 9-9).[9]

**Figure 9-9**
SS7 components are transported over IP using SIGTRAN components.

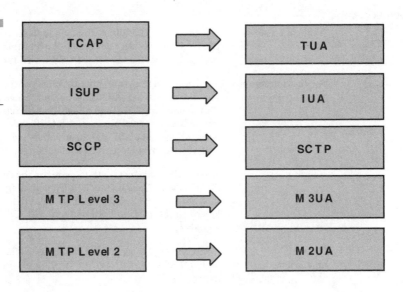

---

[8]Performance Technologies. "Signaling in Switched Circuit and VoIP Networks."

[9]Collins, pp. 337–338.

# SS7 Interworking with SIP and H.323

Until now, this chapter has described the interworking of SS7 nad IP networks. The following pages will explain how SS7 interworks with VoIP signaling protocols, SIP and H.323. As these are now the defacto standards for VoIP signaling, it is important to understand how SS7 works with each of these signaling protocols.

## ISUP Encapsulation in SIP

SCTP and the associated adaptation layers provide the capability to convey SS7 application protocol information from an endpoint in the PSTN to a call agent in the center of the VoIP network. The call agent has access to the signaling information needed to route calls on the VoIP network. SS7 application protocols themselves are not used internally in the VoIP network. Protocols such as ISUP are mapped to equivalent VoIP protocols such as SIP. SIP/ISUP mapping is the most common. This mapping works well for those calls where the call originates on a VoIP network and terminates on the PSTN. Difficulties arise when the call originates on the PSTN and terminates on the PSTN.

Rarely do the messages and parameters of one protocol match those of another. Interworking has to address the differences between state machines, times, and so on. For example, ISUP provides information that does not easily map to SIP headers. To accommodate SIP-ISUP interworking, ISUP information is mapped to the closest SIP equivalent (or discarded). To accommodate the transport of ISUP over an SIP-enabled network, SIP enables the message body to encapsulate an ISUP message. SIP messages and responses are used between the gateways, and those messages must contain an SDP message body in order for the gateways to describe media sent between them. Some messages can contain an ISUP message in binary form within the message body. The message body can have multiple parts, an SDP session description used by the gateways, and an encapsulated ISUP message carried transparently across the network (see Figure 9-10).[10]

---

[10]Ibid., pp. 350–351.

**Figure 9-10**
SIP in a transit
network can be
nontransparent to
ISUP signaling.

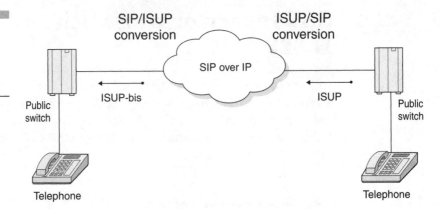

SIP provides its own reliable transfer mechanism independent of the packet layer. For this reason, SIP does not require the services of the SIG-TRAN SCTP protocol and functions reliably over an "unreliable" datagram protocol like UDP.

## SIP for Telephones (SIP-T)

*SIP for telephones* (SIP-T) is a mechanism that allows SIP to be used for ISUP call setup between SS7-based public switched telephone networks and SIP-based IP telephony networks. SIP-T carries an ISUP message payload in the body of an SIP message. The SIP header carries translated ISUP routing information. SIP-T also specifies the use of the SIP INFO method for effecting in-call ISUP signaling in IP networks.[11]

## PINT and SPIRITS

SIGTRAN is not the only IETF Working Group protocol involved in defining new protocols to enable the integration of the PSTN with IP networks. *PSTN and Internet Interworking* (PINT) and *Service in the PSTN/IN Requesting Internet Service* (SPIRITS) are two IETF Working Group recommendations that address the need to interwork telephony services between the PSTN and the Internet. PINT deals with services originating in an IP network; SPIRITS deals with services originating from the PSTN.

---

[11]Performance Technologies. "Signaling in Switched Circuit and VoIP Networks."

In PINT, PSTN network services are triggered by IP requests. An SIP Java client embedded in a Java servlet on a web server launches requests to initiate voice calls on the PSTN. The current focus of this initiative is to allow web access to voice content and enable click-to-dial/fax services. In SPIRITS, IP network services are triggered by PSTN requests. SPIRITS is primarily concerned with Internet call-waiting, Internet caller-ID delivery, and Internet call-forwarding.[12]

## Interworking H.323 and SS7

Interworking between an H.323 architecture and SS7 could be similar to interworking between softswitch and SS7 (see Figure 9-11). The approach would be to use signaling gateways that terminate SS7 signaling. The SS7 application information would be carried by using SIGTRAN from the SS7 gateways to one or more H.323 gateways.

The gateway itself can terminate SS7 links directly. In the H.323 architecture, the gateway itself contains a great deal of the application logic required for call setup, and the gateway is also the entity that performs the media conversion. By holding the application logic, it must send and receive

**Figure 9-11**
Interworking H.323
and SS7

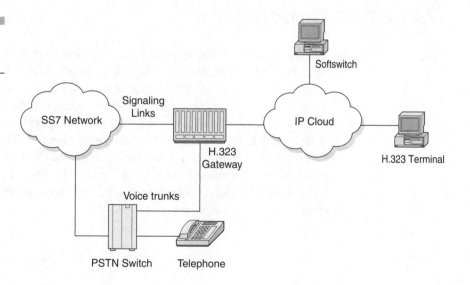

---

[12]Ibid.

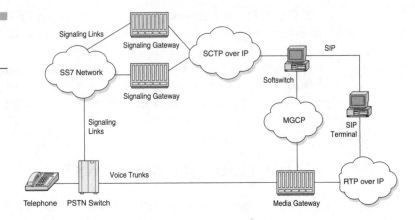

call control signaling information. As the gateway performs the media conversion, it should be placed at the edge of the IP network. Because the gateway is at the edge of the network, it is likely to be easier and more efficient for the gateway itself to terminate SS7 links from the PSTN directly. This architecture is shown in Figure 9-12.

# Conclusion

A major misperception regarding VoIP networks is that they don't support SS7. As explored in detail in this chapter, SIGTRAN and other functions facilitate the transport of SS7 over IP networks. The advantages of this are many. Most importantly, the 3,500 features (assuming 3,500 features were truly necessary) of a Class 5 switch can be transported over a softswitched VoIP network. It also allows the interconnection of VoIP networks with the PSTN, thus making any telephone in the world reachable from either network. Interworking facilitates the migration from *Time Division Multiplexing* (TDM) to VoIP networks by coopting existing circuit-switched infrastructure into IP networks.

# Features and Applications: "It's the Infrastructure, Stupid!"

One objection many service providers have regarding *Voice over IP* (VoIP) and softswitch technologies is the perception that they do not duplicate the 3,500 features of a Class 5 switch. The first question many would have is, what are those features by name and what functions do they perform? A consensus in the industry is that a residential customer may use a maximum of four or five features. A business customer may use upwards of a dozen features. Centrex 21 offers 21 of the most often used features. *Private branch exchanges* (PBXs) do not offer 3,500 features, as no such demand exists in the enterprise market. Subscribers access those features via their *dual-tone multifrequency* (DTMF) dial pads on their telephone handsets. Although in U.S. markets some regulatory agencies may require some features beyond Centrex 21, E911, and the *Communications Assistance for Law Enforcement Act* (CALEA), the race by Class 5 vendors to offer more and more features can be described in terms of Christensen's *Innovator's Dilemma* as "technology over abundance."

Softswitches that replace a Class 4 switch, for example, transmit features over the *Internet Protocol* (IP) network. Softswitches that replace a Class 5 switch utilize application and media servers that replicate the features found on a Class 5 switch. They potentially offer features not found on a Class 5 or not even possible with a Class 5 switch. These features are usually written in text-based languages using open standards. It is possible that, given the flexibility in creating new features, some softswitch solutions that replace Class 5 switches may eventually offer *more* than 3,500 features. If 3,500 features were a buying criteria for some service providers, then those same buyers will find softswitch very attractive. This chapter will provide an overview of how those features are delivered over a softswitch network. This chapter will also address the implication of new features in the market.

# Features in the PSTN

*Custom Local Area Signaling Service* (CLASS) features are basic services available in each *local access and transport area* (LATA). The features and the services they offer are a function of Class 5 switches and *Signaling System 7* (SS7) networks. The Class 4 switch offers no features of its own. It transmits the features of the Class 5. With almost three decades of development, the Class 4 switch has a well-established history of seamless inter-operability with the features offered by the Class 5 and SS7 networks. The features often enable service provider systems to offer high margins that result in stronger revenue streams.

Examples of features offered through a Class switch can be grouped into two major categories: basic and enhanced services. Examples of basic services include 1+, 800/900 service, travel cards, account codes, *personal identification numbers* (PINs), operator access, speed dialing, hotline service, *automatic number identification* (ANI) screening, *virtual private networks* (VPNs), calling cards, and call-detail recording.

Enhanced services include information database services (NXX number service, authorization codes, calling card authorization, debit/prepaid card), routing and screening (includes CIC routing, time-of-day screening, ANI screening, class-of-service screening), enterprise networks, data and video services (dedicated access lines, *Integrated Services Digital Network* [ISDN] *primary rate interface* [PRI] services, dialable wideband services, and switched 56 Kpbs), and multiple dialing plans (full 10-digit routing, 7-digit VPN routing, 15-digit international dialing, speed dialing, and hotline dialing). Most of these features have been standard on Class switches for many years.

The previous long list of features is evidence of the importance of features in the legacy market in which they were developed. Service providers will not give up these features and the margins they generate. In the converging market, features are equally important to reliability as service providers don't want to offer fewer features to their customers, and they will want to continue to offer high-margin features.

A softswitch solution emphasizes open standards as opposed to the legacy Class 4 or 5 switch that historically offered a proprietary and closed environment. A carrier was a "Nortel shop" or a "Lucent shop." No components (hardware or software) from one vendor would be compatible with products from another vendor. Any application or feature on a DMS-250 (Nortel's Class 4 switch), for example, must be a Nortel product or specifically approved by Nortel after a lengthy and rigorous testing period. New features could come only from the vendor and would usually take a minimum of 18 months for development and delivery. They would also often cost in the millions of dollars. Softswitch-open standards are aimed at freeing service providers from vendor dependence and the long and expensive service development cycles of legacy switch manufacturers.

# Features and Signaling

Features are a function of the Class 5 switch and the SS7 network. So how are the ANI features of the *Public Switched Telephone Network* (PSTN) transferred to a converging market where the softswitch competes with the Class 4 or 5 switch?

Firstly, SS7 information generated in the PSTN must be transported transparently across the IP network. The preceding chapter described SS7 over IP. It is possible with mechanisms that make it easier for service providers to quickly roll out new, high-margin services that the voice market will shift in favor of the service provider that can deliver those features quickly. It should be noted here that softswitch could enable a service provider to offer expanded features that are made available via the subscriber's PC or IP handset. The PC offers greater flexibility than a telephone handset in the range of communications between the subscriber and the switch. For example, *11 or any other handset input is limited relative to all that can be offered on a web page. In theory, if the 3,500 or so Class 5 features were cataloged on a web site and made more convenient for a subscriber to use, the service provider could better capitalize on those features. To date, Class 4 or 5 switches offer only the telephone handset as a user interface to features.

The *service creation environment* (SCE) of softswitch is almost unlimited and has the potential to change much of the switch market in favor of softswitch. Softswitch has the capability to relay existing features of the Class 5 and SS7 networks. In addition, as Figure 10-1 illustrates, it can offer a variety of new features, potentially leading to a total count of features in excess of the 3,500 features of the Class 5 switch.

## The Intelligent Network/Advanced Intelligent Network (IN/AIN)

How are features delivered? In the PSTN, features are made possible by the *Intelligent Network / Advanced Intelligent Network* (IN/AIN) and SS7. The IN/AIN is a telephone network architecture that separates service logic from switching equipment, enabling new services to be added without having to redesign switches to support new services.

Two concepts are contained in the IN/AIN. In the first, the network controls the routing of calls within it from moment to moment based on some criteria other than finding a path through the network for the call based on

**Figure 10-1**

A potential growth in
features with
softswitch
architecture

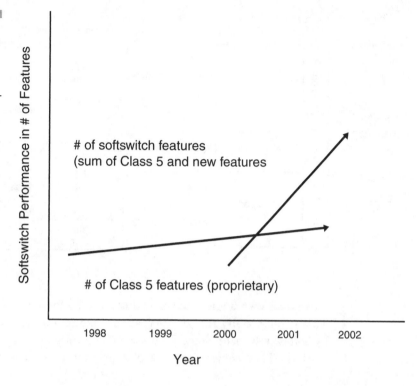

the dialed number. In the second, the originator or receiver of the call can inject intelligence into the network and affect the flow of the call. That intelligence is provided through the use of databases in a network.

The foundation of the IN/AIN architecture is SS7. IN/AIN technology uses an embedded base of stored, program-controlled switching systems and the SS7 network. At the same time, IN/AIN technology enables the separation of service-specific functions and data from other network resources. This feature reduces the dependency on switching system vendors for software developments and delivery schedules. In theory, service providers have more freedom to create and customize services.[1]

The promises of IN/AIN have yet to be fully realized in the PSTN. Incumbent telecommunications vendors have pushed their own proprietary systems that then serve to stultify innovation. Softswitch offers the advantage of enabling a service provider to integrate third-party applications or even

---

[1]Telcordia Technologies. "Intelligent Networks (IN)." A white paper hosted by the International Engineering Consortium at www.iec.org, pp. 6–8.

**Figure 10-2**
Three-layer softswitch architecture compared with legacy circuit switching. Note the open standards for interfacing between call control and switching and applications and service layers.

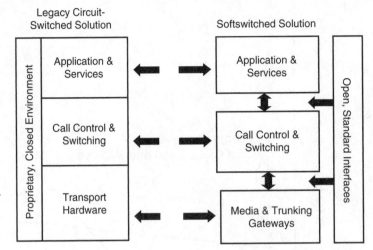

write their own while interoperating with the features of the PSTN via SS7. This is potentially the greatest advantage to a service provider presented by softswitch technology. Chapter 2, "The Public Switched Telephone Network (PSTN)," covers the IN/AIN in greater detail (see Figure 10-2).

## Service Creation Environment

The simple transport of long-distance voice is highly commoditized and offers low margins to service providers. Without services, next-generation switches would not be able to generate the voice revenue that currently provides 80 percent of the overall service provider revenue.[2] The key to high margins in the converging market is rapid service creation. The service providers that can differentiate themselves from the competition with unique services will win market share and profitable revenue margins from those services. In the legacy market, the offering of such services has been dictated by the switch vendors (no more than two to three in the North American market). The switch vendors have had little incentive to enable their customers (the long-distance carriers) to roll out new services in a rapid and flexible manner.

---

[2]Telica. "Accelerating the Deployment of Voice over IP (VoIP) and Voice over ATM (VoATM)." A white paper posted by International Engineering Consortium at www.iec.org.

Softswitch changes this scenario. By virtue of its open standards interface, softswitch enables service providers to quickly create and roll out new services. The structure of softswitch architecture makes it possible for service providers to integrate applications from the softswitch vendor or third-party vendors, or even develop their own in house. The inclusion of the option to write custom *application programming interfaces* (APIs) on a softswitch enables a service provider to write only those custom APIs that are applicable to a given service at a given location. In addition to interoperability between gateways and softswitches, the APIs are standardized to enable any third-party developer the ability to create applications on top of the softswitch (see Figure 10-3).

Service creation enables the development of new service building blocks and the assembly of services from these building blocks, typically using one or more commercially available, off-the-shelf tools, such as an *integrated development environment* (IDE). The concept here is modularity, where a service provider can mix and match components in their network regardless of the vendors involved. If the voice mail platform of a next-generation service provider is not working to the satisfaction of that service provider, it can be replaced by a competing voice mail platform. This also holds true for media gateways, signaling gateways, and so on. Gone are the days of monolithic, single-vendor networks. Next-generation network services will

**Figure 10-3**
Softswitch architecture detailing the applications and services layer as well as APIs. Given open standards of softswitch architecture, any number of third-party applications can be added to the network. Note the applications on the uppermost layer.

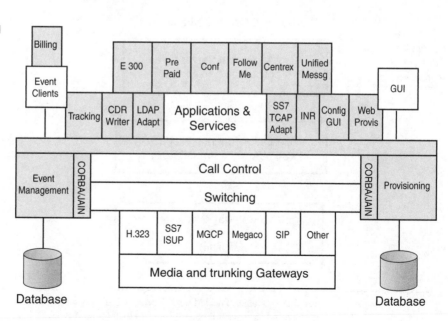

be assembled on the fly in a plug-and-play fashion, drastically reducing the time and effort to develop services.[3]

## Application Programming Interfaces (APIs)

Features reside at the application layer in softswitch architecture. The interface between the call control layer and specific applications is the API. Writing and interfacing an application with the rest of the softswitch architecture occurs in the service creation environment. Those open standards include APIs known as Parlay, *Java Advanced Intelligent Network* (JAIN), *Common Object Request Broker Architecture* (CORBA), and *Extensible Markup Language* (XML). Figure 10-4 details the relationship of APIs in the service creation environment.

## APIs and Services

To compete with the incumbent service providers, packet telephony service providers must be able to provide existing services while introducing advanced services quickly and inexpensively. By moving the responsibility

**Figure 10-4**

The relationship of APIs in the service creation environment

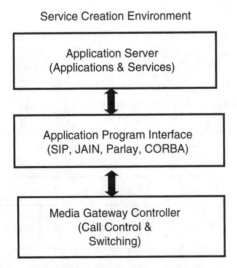

Service Creation Environment

[3]Sun Microsystems. "The JAIN APIs: Integrated Network APIs for the Java Platform." A white paper available at www.java.sun.com/products/jain, May 2002, pg. 7.

for advanced services away from the call control entities, APIs resident on an application server, such as JAIN, Parlay, CORBA, *Call Processing Language* (CPL), *Common Gateway Interface* (CGI), and Servlets, provide packet telephony service providers with an environment where differentiating services can be deployed rapidly. The open (as opposed to proprietary) nature of these APIs enables services to be created, managed, and deployed without requiring new or upgraded network infrastructure functionality.

One of the major advantages of using packet telephony and softswitch architecture is that the underlying transport protocol (IP) is also used for a variety of other services, including web access, email, presence, and instant messaging. Enhanced services in packet telephony networks will enable the integration of these services, bringing together telephony and Internet services. Application servers are required to interwork with a variety of non-telephony protocols and APIs in order to provide these services. Because of the complexity and overhead, service APIs reside on an application server to provide enhanced services processing. Services on an application server can provide the integration of traditional AIN/IN services with Internet services.

The application server model provides an architecture for enhanced services within a pure packet-based network and within the business domain of the network provider. APIs such as Parlay, JAIN, and CORBA focus on hybrid networks and provide for services both inside and outside the network provider's business and technology domains. Third-party developer relationships are extremely important to softswitch vendors. Assuming media gateway ports become commoditized, softswitch vendors will need to distinguish themselves by offering the easy deployment of third-party applications. Service providers will value third-party developers, which can create the applications that will set carriers apart from their competitors.[4] Firms exist now that own no switches or network and provide only application services to carriers. Open standards enable a carrier to "shop" the market for components that are more competitively priced than what one would find when buying proprietary components from the Class 4 vendor, for example.

**Parlay**   Parlay is a platform for Java technology-based enterprise application integration, development, and deployment. Its features and enterprise scalability combine to solve two key enterprise business challenges: integrating diverse types of data, applications, and client environments,

---

[4]Pfalmer, Brandy. "Will Softswitch Be the Heart of Next-Generation Networks? Call State and Interoperability Issues Continue to Hamper Softswitch Beat."*Sounding Board,* October 1999, pg. 15.

and facilitating the rapid application development via distributed and scalable Java technology-based components.

Parlay's business benefits to large enterprises, software developers, and partners are compelling. They let organizations leverage investments in existing business applications and deliver time-to-volume and time-to-market advantages of developing applications on a component-based architecture. Parlay also reduces administrative and management costs of the resulting applications, including business solutions that are easily deployed in softswitch environments.[5] The Parlay Group, a consortium of software vendors, controls the standards for Parlay.

**JAIN Network Topology**   Most, but not all, softswitches are now hosted by servers from Sun Microsystems. JAIN is the API of choice by Sun and its partners. By providing a new level of abstraction and associated Java interfaces for service creation across PSTN, IP, or ATM, JAIN technology enables the integration of IP and IN protocols. By allowing Java applications to have secure access to resources inside the network, the opportunity is created to deliver thousands of features rather than the dozens currently available. Thus, JAIN technology is changing the telecommunications market from many proprietary closed systems to a single network architecture where features can be rapidly created and deployed.

JAIN APIs are a set of Java-based APIs that enable the rapid development of telecom features on the Java platform. The JAIN APIs bring service portability, convergence, and secure network access to telephony and data networks.[6] Table 10-1 outlines the goals of the JAIN APIs.

The JAIN network topology provides carriers with the ability to deploy next-generation network services on devices inside or at the edge of the network, including any Java-enabled end-user device. Furthermore, support for all the necessary telephony protocols that are used between the different network elements in IN-, AIN-, and IP-based (telephony) networks is mandatory.

A key aspect of the JAIN component architecture is to move the signaling layer away from proprietary switches into open call control servers or softswitches. Signaling, that is, the protocols used to establish and terminate communication connections, is the common thread between conventional telecommunications switches and softswitches. The ability to adapt

---

[5]Information Builders, Inc. "Parlay Application Server." Published by Sun Microsystems, Inc. at http://industry.java.sun.com/solutions/products/by_product/0,2348,all-2198-99,00.html.

[6]Sun Microsystems. "The JAIN APIs." http://java.sun.com/products/jain/overview.html.

**Table 10-1**

Goals of the JAIN Initiative

| Goal | Description |
|------|-------------|
| Service portability | Write once, run anywhere. Technology development is currently constrained by proprietary interfaces. This increases development cost, time to market, and maintenance requirements. With the JAIN initiative, proprietary interfaces are reshaped to uniform Java interfaces delivering truly portable applications. |
| Network convergence (integrated networks) | Any network. By delivering the facility to enable applications and services to run on PSTN, packet, and wireless networks, JAIN technology speeds network convergence. As the demand for services over IP rises, new economies of scale are possible as well as more efficient management and greater integration with IT. |
| Secure network access | By anyone. By enabling applications residing outside the network to directly access network resources and devices to carry out specific actions or functions, a new environment is created for developers and users. The market opportunity for new services is huge when controlled access is provided to the available functionality and intelligence inside the telecommunications networks. |

Source: Sun Microsystems[7]

the signaling components between networks is paramount to the success of carriers and network service providers.

Figure 10-5 is a representation of when JAIN APIs are defined within a communications platform. The softswitch architecture is centered on mapping the call control/session interfaces onto the underlying protocol APIs. Since softswitches perform signaling on IP networks, most are equipped with the *Session Initiation Protocol* (SIP), the *Media Gateway Control Protocol* (MGCP), MEGACO, or H.323 underlying protocols. Several softswitches also include SS7 protocols to address interfaces for the existing telephone network.[8]

**Common Object Request Broker Architecture (CORBA)** CORBA, is the Object Management Group's (an industry standards group comprised

---

[7]Ibid.

[8]Ibid., pg. 10.

**Figure 10-5**
JAIN architecture

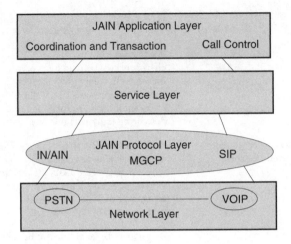

largely of software vendors) open, vendor-independent architecture and infrastructure that computer applications use to work together over computer networks. Using the standard protocol *Internet Inter-ORB Protocol* (IIOP), a CORBA-based program from any vendor can interoperate with a CORBA-based program from the same or another vendor, on almost any other computer, operating system, programming language, and network.[9]

**Extensible Markup Language (XML)**   XML, the next generation of HTML, is now viewed as the standard way information will be exchanged in environments that do not share common platforms. XML 1.0 was released in February 1998. XML-based network management uses the *World Wide Web Consortium's* (W3C) XML to encode communications data, providing an excellent mechanism for transmitting the complex data that is used to manage networking gear. Building an API around an XML-based *remote procedure call* (RPC) gives a simple, extensible way to exchange this data with a device. Receiving data in XML opens options for handling the data using standards-based tools. XML is widely accepted in other problem domains, and free and commercial tools are emerging at an accelerating rate.[10]

[9]Object Management Group. "CORBA Basics." www.omg.org/gettingstarted/corbafaq.htm# WhatIsIt.

[10]Shafer, Phil. "XML-Based Network Management." A white paper from Juniper Networks, August 2001. Available at www.juniper.net/techcenter/techpapers/200017.html

**Microsoft's XML .NET Architecture**    As mentioned in Chapter 5, "SIP: Alternative Softswitch Architecture," Microsoft is entering the telecommunications market by including SIP in its Windows XP system. In other strategies, Microsoft now offers its .NET architecture, which includes XML and *Real-Time Communications* (RTC). The RTC features enable programmers to build applications that do text messaging, VoIP, bridging from IP to PSTN telephony, and receiving/sending presence information. RTC APIs are built on SIP. SIP enables a device to call, or receive calls from, any SIP client. A SIP client can be a Windows CE device, Windows XP machine, or a third-party SIP *user agent* (UA). The Windows Messenger application supports RTC as a backend.

RTC sets up VoIP connections between Windows CE .NET devices, other Windows CE devices, Windows XP systems, third-party SIP UAs, or PSTN phones. It initiates text-messaging sessions, which can occur simultaneously with voice connections. RTC also receives presence information to see if user contacts are online, it enables users to change their own presence state and send notifications, and it monitors who is watching the user's online presence.

A PCI bus is reserved for an appropriate audio card. If the audio is on board the PC, it must have an AC97 *coder/decoder* (codec) along with a codec controller that provides a hardware-assisted audio data transfer to and from the codec. The controller must have enough bandwidth to service the audio and other devices in the system, such as a network *integrated development environment* (IDE) and modem.

The codec controller must also have an interrupt mechanism that can notify the driver when a data transfer is complete. Additionally, the controller should contain sufficient buffering capabilities in audio data to enable the interrupt service thread to start another data transfer without causing audio breakup.

To support VoIP, the capture device and driver should be able to handle wave-in-capture buffers as small as 10 milliseconds and return them to the application when they are filled with data. Note that peripheral hardware like the microphone and earpiece should ideally have automatic gain control and built-in echo cancellation.[11] However, new software from Microsoft includes volume control and echo cancellation.

---

[11]Microsoft Corporation. "Real-Time Communications Overview." A white paper available at http://msdn.microsoft.com/library/default.asp?url=/library/en-us/wceoak40/htm/coconrtcoverview.asp.

# SIP as API: International Softswitch Consortium's Architecture for Enhanced Services in a Softswitch Network

The International Softswitch Consortium (see www.softswitch.org), an industry group of softswitch vendors and service providers, has proposed an architecture for the delivery of features on softswitch networks. Although this industry is still in its infancy, the ISC proposes certain guidelines aimed at promoting open standards and open architecture. The following section outlines the architecture for enhanced services proposed by the ISC. Two components, the application server and the media server, are introduced into the architecture to provide support for enhanced service logic, management functions, and specialized media resources. Application servers are intended to host a variety of enhanced services. An application server provides a framework for the execution and management of enhanced services. These services make use of the call control functions provided by the underlying infrastructure as well as enabling the integration of messaging, presence, and web services (see Figure 10-6).

**Figure 10-6**
The functions of application and media servers (Source: International Softswitch Consortium)

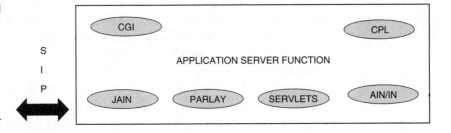

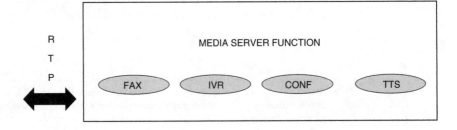

## Media Servers

A media server provides specialized resources such as *Interactive Voice Response* (IVR), conferencing, and facsimile functions. Media servers and application servers are independent and can be deployed on separate physical platforms or on the same platform. An application server can utilize resources on a media server for enhanced services, which require access to the media stream.

## Application Servers

APIs, as described earlier in this chapter, are hosted on the application server, providing access to underlying service and switching functions. Using these APIs, services can be easily developed and deployed. In addition, traditional telephony call models and *Transaction Capabilities Application Part/Intelligent Network Application Part* (TCAP/INAP) protocols can potentially reside on an application server, providing access to standard AIN/IN telephony services.

## Architecture

A functional view of the next-generation packet telephony architecture is shown in Figure 10-7. An example of SIP control flow between application server and call control functions is displayed in Figure 10-8. The function of each entity in Figure 10-8 is described in Table 10-2.

The introduction of an application server to the packet architecture provides for platforms specifically designed for enhanced services. By using SIP as the application interface between a call control entity (softswitch) and an application server, a common, standard protocol can be used for communication between all call control and service execution entities. These new services can be deployed quickly with little or no impact on network rollouts.

An application server utilizes an SIP interface, which provides access to the signaling aspects of a call, while a media server utilizes a *Real-Time Protocol* (RTP) interface, which provides access to the media stream. Services on application servers make use of the resources provided by media servers. With an application server and a media server, services that require access to both the signaling path and media stream can be provided.

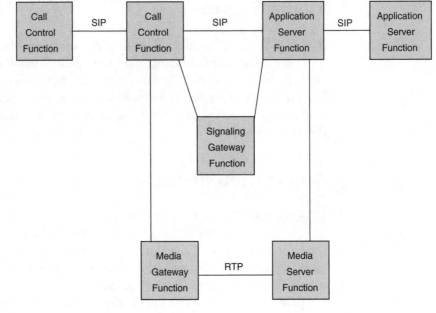

**Figure 10-7**
Softswitch-enhanced service architecture (Source: International Softswitch Consortium)

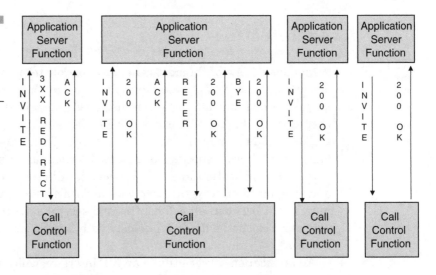

**Figure 10-8**
SIP control flow between application server and call control functions

Enhanced services reside on application servers, which provide an environment for the execution, management, and creation of these services. Service APIs, as well as AIN/IN call models, may reside on the application server. To support converged services, an application server may also utilize

**Table 10-2**

Functions of
enhanced services
architecture entities

| Function | Description |
|---|---|
| Call control function | May provide connection control, translations and routing, gateway management, call control, bandwidth management, signaling, provisioning, security, and call detail record generation. |
| Media gateway function | May provide a conversion between circuit-switched resources (lines and trunks) and the packet network (IP and ATM), including voice compression, fax relay, echo cancellation, and digit detection. |
| Signaling gateway function | May provide a conversion between the SS7 signaling network (SS7 links) and the packet network, including protocols such as ISUP and TCAP. |
| Application server function | May provide for the execution and management of enhanced services, handling the signaling interface to a call control function. It also provides APIs for creating and deploying services. |
| Media server function | May provide for specialized media resources (IVR, conferencing, facsimile, announcements, and speech recognition) and handling the bearer interface to a media gateway function. |

protocols and APIs for functions such as email, instant messaging, presence, web access, and so on.

## Physical Architecture

This section discusses two possible physical architecture models, showing how the functional architecture maps to a "real" physical architecture. This section is used to provide a better understanding of how the logical architecture is actually implemented. Two physical architectures are discussed: a centralized architecture and a distributed architecture.

**Centralized Architecture** Figure 10-9 shows a centralized physical architecture. The softswitch, in this example, provides for call control, signaling, and basic application functions. The application server function in the softswitch provides for standard services, such as call waiting or calling line identity, that are not typically provided on external platforms. The media gateway provides an interface to circuit-switched networks. The

**Figure 10-9**
A centralized
architecture for
enhanced services
(Source: International
Softswitch
Consortium)

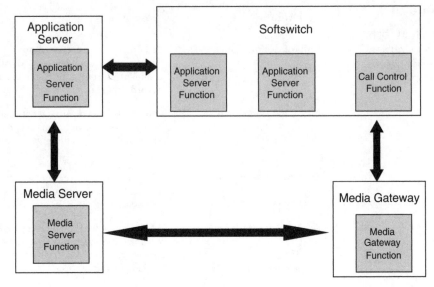

application server and media server map directly from the functional architecture. The softswitch can communicate with other application servers and softswitches using SIP.

**Distributed Architecture**    Figure 10-10 shows a distributed physical architecture. The mediation gateway functions as a protocol converter between circuit-switched and packet networks, providing media gateway, signaling gateway, and basic call control functions. Functionality-wise, an IP phone or customer premise gateway is identical to a large-scale gateway. Both use SIP as the interface with other call control and service entities.

In this architecture, a feature server is used to provide service-level routing between call control entities and service entities. The application server in this example can provide services based on the signaling path, while the media server can provide media-based services. All entities in this architecture communicate via SIP, providing a common peer-to-peer interface.

## Interface Between Call Control and Application Servers

SIP is used as the interface between call control entities (softswitch) and application servers because of its general acceptance, availability, and capa-

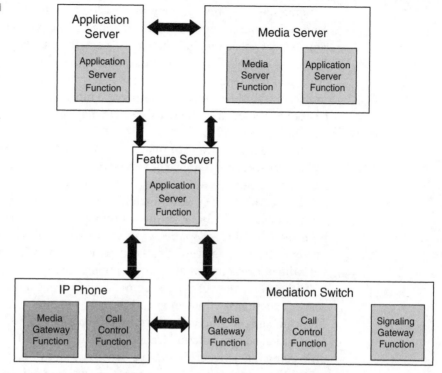

**Figure 10-10**
A distributed physical architecture for enhanced services (Source: International Softswitch Consortium)

bility to set up, tear down, and manage sessions between end-points. In this context, a call control entity has the capability to set up and take down a call-signaling path to an application server, and for an application server to set up and take down a call-signaling path to a call control entity. It also includes the capability to convey calling and called party information, hold and resume connections, transfer sessions, and establish multiparty connections. SIP is used only for signaling; RTP is used to carry the media. SIP relays the information necessary to establish RTP communication between end-points.

When used in conjunction with a media server, an application server can provide a rich set of enhanced services. The service logic in the application server has access to all call events via SIP signaling. An application server interacts with a media server for access to the media stream to detect DTMF digits, play and record media, mix media, detect and relay fax transmissions, play announcements and tones, and perform speech recognition. This multimedia access provides support for complex services such as unified messaging, conferencing, calling/debit card, and call-center-type applications.

Some media gateways may also include media server functions. If the required resources are available on a media gateway, an application server can request these media functions via a call control entity. Typically, these functions include DTMF collection and generation, call progress tone generation and detection, and audio playback and recording. Media processing can occur at either a media gateway or a media server. Gateways are always located at the edge of the network, but media servers can be located at the edge or in the core.

The reader may want to refer back to Chapter 5 to review SIP. Call control entities, such as gateways, softswitches, and IP phones, act as SIP UAs. Application servers may act as SIP entities: UAs, redirect servers, proxy servers, or third-party call controllers (back-to-back UAs). All entities may communicate directly or through proxy servers. A register mechanism is required to enable an application server to inform a call control entity of its availability. Alternatively, a call control entity may be statically configured with the address information of an application server.

The general flow of events from a call control function to an application server function is illustrated in Figure 10-11 and is described as follows. The call control function determines that a call is to be handed off to an application server function (that is, send an INVITE) for enhanced services processing. The triggering may be based on the calling party, called party, or some other mechanism. The call control function determines the address of

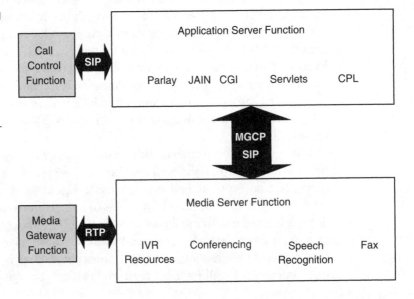

**Figure 10-11**
Interfacing application and media servers (Source: International Softswitch Consortium)

the application server function based on the trigger information and relays the call to the application server function, including the appropriate call information. The application server function receives the call and invokes the corresponding enhanced service. Next, the application server function can perform any of the actions in Table 10-3.

In addition to the previous actions, an application server may initiate calls based on some other input, such as a web action, email, presence notification, or instant message. This enables an application server to support third-party call control. An application server provides an interface to the user via the web, email, presence, instant message, and so on. A user can then initiate actions that are received by the application server, which, acting on behalf of a user, initiates the actions using SIP. This capability provides for support of web or message-based call control, enabling powerful and exciting applications.

Basic SIP, in conjunction with call control extensions, can provide a mechanism for enabling a call control function to hand off calls to an application server for call control and enhanced services processing. An application server can optionally request access to the media stream, interacting with a media server. After processing, an application server can transfer a call back through a call control entity and disconnect itself from the call. This provides a powerful mechanism for enabling new and exciting enhanced services, including services that require access to the media path.

**Table 10-3**

SIP control flow for interfacing between a call control entity and an application server function

| SIP function | Description |
|---|---|
| Redirection | Redirects the call by sending a new destination address back to the call control function. This mechanism can be used to implement translation- and routing-oriented services. |
| Accept and transfer | Assigns a media resource from a media server function and orders the call control function to connect the media path to it. After the user interaction with the media resource is complete, the application can transfer the call to a new destination and drop out of the call. This mechanism can be used to implement media-oriented services, such as calling card and fax store/forward services. |
| Proxy | Sends the call back through a call control function. This enables the application to monitor all subsequent call events. This mechanism can be used to implement event-oriented services, such as debit card and timing services. |

Source: International Softswitch Consortium

## Application Server to Media Server Interface

A media server hosts specialized resources, including those for facsimile, IVR, conferencing, speech recognition, and text to speech. These resources are used by service logic in the application server to provide enhanced services requiring interaction with the media or bearer path. An application requests resources from a media server and associates them with a service instance. Then the application server requests that the RTP media stream be redirected to the selected resource. H.248/MEGACO, MGCP, or SIP (which may also require other supporting protocols) may be used as the control interface between the application server and media server. Figure 10-11 illustrates this interaction.

## Service APIs

To compete with existing, entrenched service providers, packet telephony service providers must be able to provide existing services while introducing new, advanced services quickly and inexpensively.

SSP call models for AIN/IN can be placed on an application server. This allows a call control entity to delegate AIN/IN SSP functions to an application server, which hosts the appropriate call model. With this approach, an application server can interact with AIN/IN SCPs and IPs to support legacy services and platforms. The interface between a softswitch and application server remains unchanged with no impact on the call control entities.

Application server and call control functions can access existing AIN/IN entities. As mentioned earlier, services on an application server can integrate traditional AIN/IN services with Internet services. Using this approach, existing circuit-switch service providers can leverage their existing platforms while keeping pace with packet telephony providers. Circuit-switch providers can also choose to leave existing networks in place while independently deploying next-generation packet telephony networks to support advanced converged services (see Figure 10-12).

## Application Server Interactions

In addition to using SIP as the interface between application servers and call control entities, it can also be used as the interface between application servers. This allows application servers to interact, providing the capability

**Figure 10-12**
Interaction of IN
and SIP (Source:
International
Softswitch
Consortium)

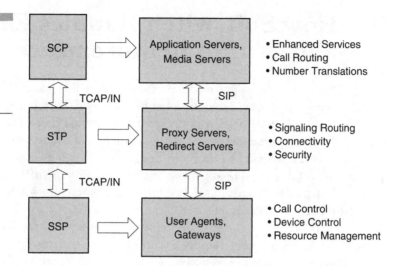

for two or more enhanced services on the same or different application servers to be linked together. With the capability of an application server to delegate control to other application servers, a mechanism of managing feature interactions can be introduced. This becomes critical as more enhanced services are introduced into a network.

## Application and Media Servers Summary

The introduction of application and media servers into next-generation packet telephony networks provides a vehicle for the deployment of enhanced services. Utilizing a standard protocol between call control entities and application servers, enhanced services can be quickly introduced on application servers by application experts. Media servers can be used to provide specialized bearer resources required by many enhanced services. Standardized service APIs and AIN/IN call models, which reside on an application server, can be used to enable service developers access to underlying telephony network functions. In addition, application servers can also utilize Internet protocols and APIs to provide truly converged services.[12]

---

[12]Hoffpauir, Scott. "Enhanced Services Framework." A white paper for the International Softswitch Consortium.

# How Softswitch Handles E911 and CALEA Requirements

No two features get more discussion than E911 and CALEA. Incumbent service providers have invested millions of dollars in their infrastructures to comply with legal requirements to provide these features. When VoIP switches first came on the market, vendors of circuit switches used the lack of E911 and CALEA in those fledgeling devices as a means to spread fear, uncertainty, and doubt regarding the viability of VoIP and VoIP platforms. Much has changed in the last few years. Some softswitch vendors now offer E911 and CALEA features.

It should be stated here that only primary line service providers in the United States are required to offer these features. Long-distance service providers are not required to offer these services. Cell phone service providers are not required to provide E911. An internet subscriber is not required by law to subscribe to E911 or CALEA when making VoIP calls from their computer. Most service providers outside the United States are not required to provide E911 or CALEA. Softswitch vendors should note that perhaps a majority of the world's voice traffic travels on networks that have no requirement for E911 and CALEA.

## Enhanced 911 (E911)

One obstacle to the wide deployment of softswitch is accommodating the demands of *Enhanced 911* (E911) service. Although 911 services enable users to quickly request service from emergency personnel, E911 goes a step further by using ANI to relate location information to that number and determine which *public service answering point* (PSAP) should handle the call.

In an enterprise setting, it is not only important for E911 to direct emergency services to a particular building, but to indicate from where in the building the 911 call originated. Typically, when someone dials 911 from their desk, the ANI information that reaches the PSAP is based on the company's general, seven-digit phone number associated with the PBX, not the user's extension. In a life-and-death circumstance, that incorrect information can create harmful delays. As a result, some local and state governments make incorporation of E911 services a legal obligation. To that end, many private phone system vendors have incorporated ways for PBXs to fulfill these E911 requirements.

Different vendors offer different strategies toward accommodating the demands of E911. The most straightforward way is to relate circuit terminations to an extension number. In a traditional PBX environment, each phone jack is associated with a particular extension. When a 911 call is placed from a particular location, the PBX sends a seven-digit number associated with that extension to a central office router, which can then use ANI to route the call to the correct PSAP. The PSAP can then furnish emergency personnel with that special number's location information.

However, in a packetized network, the circuit-switched world's E911 fix doesn't apply to SIP (or any other VoIP protocol). Since SIP phones and other VoIP devices become "floating" network appliances that can attach to the network from any point, their location isn't static. Just like a laptop, a SIP phone can connect anywhere on a company's network; all it needs is a valid IP address. If the phone can be anywhere, then how can a company's phone system serve the E911 network with the necessary information?

Several fixes to this problem have arisen from within the SIP community:

- Install *Global Positioning System* (GPS) chipsets in SIP end-user devices that can provide geographic location information.

- Simply have users log on to their devices when moving them from one location to another.

- If an SIP deployment makes use of preexisting voice cabling to create a second, "VoIP-only network, the IP PBX can relate circuit terminations to *Direct Inward Dialing* (DID) lines in much the same way as a circuit-switched PBX.

Whatever the solution, fixes to the E911 issue will have to be standardized across SIP products from all vendors since the E911 space is regulated.[13]

# Communications Assistance for Law Enforcement Act (CALEA)

CALEA is a U.S. law providing for the wiretapping (intercept) of information from a telecommunications network. CALEA dates from the 1990s and

---

[13]Mitel Networks Corporation. "SIP at the Desktop Intelligence to the Edge." September 2001, pg. 7. A white paper from Mitel, available online at www.sipcenter.com/files/SIP_at_the_Desktop.pdf.

is not a traditional requirement of the PSTN. Another concept related to CALEA is lawful intercept, that is, when a judge has issued a court order allowing the wiretapping of a given telephone number (not a person). The court order is awarded to a law enforcement agency who in turn produces the court order to a telecommunications service provider. Multiple court orders by different law enforcement agencies may exist for a given telephone number.

Wiretaps fall into two categories: call detail and call content. In a call detail wiretap, information regarding the various calls sent and received from that phone number are compiled. That is, it records what numbers were called or from what numbers calls were received, the date and time of each call, and the duration of each call. Call content information consists of recordings of the calls themselves, revealing the actual content of the call. The suspect must not detect the wiretap. The tap must then occur within the network and not at the subscriber gateway or customer premise. The wiretap must not be detectable by any change in timing, feature availability, or operation.

In the PSTN, a lawful intercept occurs in the Class 5 switch as it avoids any contact with the subscriber gateway or customer premise. The concern in the industry is that in a VoIP network that does not have a Class 5 switch, no easy mechanism will be available for installing wiretaps.

It is possible to perform a lawful intercept in a VoIP network. VoIP networks contain separate call agents and media gateways. The call agent is responsible for all call control and is the element that collects all the details of the calls required in a call detail tap. Hence, a softswitch can provide a call detail wiretap solution. The call agent does not see the call content contained in the RTP media stream, so call content must be collected elsewhere in the network.

In the softswitch industry at the time of this writing, the deployments of Class 5 replacement softswitches are limited. Softswitch solutions that replace Class 4 switches are the most prevalent applications. In the case of Class 4 replacement softswitch solutions, the CALEA objection to softswitch technology is irrelevant because in the PSTN lawful intercept does not occur in a legacy Class 4 switch. Admittedly, assuming a progression in technology where a Class 5 replacement softswitch actually replaces a 5ESS in the PSTN, there will no doubt be a legal requirement on the part of the service provider to provide CALEA services to law enforcement agencies. It is possible that call content collection would have to take place farther out toward the periphery of the network.

As described earlier in this book, a variety of network architectures have been made possible by softswitch. CALEA applies to those service providers offering their services as a primary line. CALEA does not apply to service providers offering their voice services as a secondary line. This applies to service providers in the United States. Not all markets in the world have similar requirements. CALEA is a valid concern for a vendor wanting to sell their softswitch solution to a U.S.-based *Regional Bell Operating Company* (RBOC, of which four exist) as a Class 5 replacement. Given the variety and flexibility of softswitch solutions, compliance with CALEA is neither impossible nor as clear-cut as it is in a TDM PSTN setting.

# Next-Generation Applications Made Possible by Softswitch Features

The development of new and profitable features is what will mark the near term success of softswitch applications. It is impossible to forecast now exactly which features will later be considered "killer apps" of VoIP and softswitch. The following paragraphs speculate on a few features available at the time of this printing.

## Web Provisioning

One unique feature arriving on the market with softswitch applications that is not available on Class 4 is web provisioning. Web provisioning enables customers to do their own provisioning via a web site. This also allows customers to choose their own product mix and to turn individual services on or off. Other features available on softswitch but not on Class 4 include voicemail to email, voice email browsing, voice calendar (which calls you about events), voice-enabled dialing, flexible call routing based on external events, click to dial on PC (phone rings and connects to party), remote call control (forwards a call including feature via a web page), and web phone control. In short, by using *VoiceXML* (VMXL) programming, a new feature can be written anywhere from 20 minutes to a few days.[14]

---

[14]Interview with Nathan Stratton, CTO, Exario (www.exario.com), October 28, 2001.

## Voice-Activated Web Interface

Many cellular service providers already offer voice-activated dialing where the subscriber's voice interfaces a database of telephone numbers and selects one to be dialed. That technology can be taken a few steps further to be voice-activated web interfacing. Instead of needing a computer to interface web sites or email, a subscriber could access those resources via a voice-activated interface. In short, a subscriber could retrieve driving instructions via their cell phone or obtain stock quotes, the latest news, and weather, as well as email. SandCherry of Boulder, Colorado, is actively engaged in developing this technology for use by telephone subscribers.

## "The Big So What?" of Enhanced Features

At the time of this writing, it is still far too early to determine what the killer app is as regards applications for softswitch that cannot be duplicated on a Class 4 or 5 switch. Applications such as follow me, web provisioning, and voice-activated web pages may look promising at this time, but it should be stressed that it is not the specific application that constitutes a killer app in context, but rather the infrastructure that makes any new and convenient application possible. An open-standard, IP-based infrastructure makes a variety of new applications possible. Not only that, but using SIP or VXML to develop and deliver these applications adds a great deal of flexibility for service provider and subscriber alike. Perhaps the best definition of the new services made possible by softswitch architecture and applications is provided by Ike Elliott, Chairman of the International Softswitch Consortium, who describes them as being "that feature or set of features that makes a worker more efficient." In other words, this architecture offers a business or industry the flexibility to write features and applications that are specific to those businesses or industries. Historically, a business had to take what the equipment vendor was offering or would provide for a steep fee.

## Napster and i-Mode: Examples of Killer Apps

No one can predict what the killer app will be. A good example would be Napster, which provided MP3 downloads via the Internet free of charge. At one point, it was rumored to be about one-third of all internet traffic. This killer app was not the product of a century-old Fortune 10 service provider or even a media conglomerate. It was the product of a college dropout with a few servers connected to the Internet.

An example of the features and applications that generate a great deal of revenue for the service provider is Japanese cell phone service provider DoCoMo's i-Mode. i-Mode provides HTML-based (text-based language delivered over a packet network) information to DoCoMo's cell-phone-enabled subscribers. The DoCoMo subscribers can access graphic information from their cell phones. Some 46,000 unregulated sites can be reached by typing in a URL, but the 1,800 official i-Mode offerings are constantly monitored by DoCoMo to make sure they are current and easy to use. Official i-Mode sites are allowed to charge ¥100 to ¥300 a month ($0.85 to $2.50), which DoCoMo collects for them in exchange for a 9 percent fee. Although these sites are created with a compact version of HTML, the lingua franca of the Web, terms like HTML and even Web never appear in i-Mode ads. Handsets are manufactured by name-brand consumer electronics outfits like Sony and Panasonic, but DoCoMo subsidizes the phones heavily to keep the price low: A model that might ordinarily cost $600 retails for less than $350. Because DoCoMo's wireless network is packet-switched, users are charged only for the number of data packets they send back and forth, not for the amount of time they're connected, as on most networks outside Japan.

i-Mode, introduced with minimal expectations in February 1999, has attracted more than 25 million subscribers—one-fifth of Japan's population. New subscribers are still signing on at the rate of 43,000 a day, 1.3 million a month. i-Mode now has 39 million cell phone subscribers and generated revenues last year of $39 billion with 180 employees. That translates to $216 per employee per year. Compare this to Qwest and WorldCom at $295,000 and $229,000 respectively per employee per year (see Table 10-4).[15]

# Conclusion

Given a flexibility relative to Class 4 and 5 switches, softswitch solutions can offer just as many of the necessary features that the PSTN switches

---

[15]Rose, Frank. *Wired Magazine*, September, 2001. Available online at www.wired.com/wired/archive/9.09/docomo.html.

**Table 10-4**

Comparison of
i-Mode and leading
U.S. service
providers in terms
of revenues per
employee

| Company | Gross annual revenue per employee |
|---------|-----------------------------------|
| i-Mode | $216 million ($39 billion with 180 employees) |
| Qwest | $295,000 ($18 billion with 61,000 employees) |
| WorldCom | $229,000 ($14 billion with 61,000 employees), approximately one one-thousandth of i-Mode |

Source: 2001 revenue figures for respective firms

offer. Given the ease of writing and deploying new features relative to legacy networks, it is possible that softswitch solutions potentially offer more than 3,500 features.

Imagine an application that is web based and hosted on a small business' server. A variety of service providers could provide their subscribers with access to that application and charge the subscriber for its use. Revenues for that application are shared between the service provider, the creator of the application, and the business that hosted and maintained the application. This economic model, given what could be potentially a low barrier to entry, could spur rapid growth in new and innovative features hosted on application servers located anywhere on an IP network, generating significant revenue for software writers and the service providers that made them available to subscribers. Such a firm would not necessarily have to generate $39 billion in annual revenues with 180 employees such as i-Mode, but any fraction of that equation could prove to be a very lucrative business model. Let a thousand flowers bloom; let a hundred schools of thought contend.

# Softswitch Economics

# A Previous Example of Disruptive Technology in the Long-Distance Industry

The converging market demands greater flexibility in switching equipment. In order to take advantage of market opportunities, a service operator must be able to enter and exit markets with a degree of ease. This chapter will detail how softswitch is cheaper to purchase and operate (in terms of rental space and power draw) and is smaller in footprint (size and shape) than Class 4 or 5. The low cost of *Voice over IP* (VoIP) technology and operations relative to legacy circuit switches is one reason why softswitch can be considered a disruptive technology.

The reasons for the success of VoIP gateway switches in the long-distance bypass market were that the switches were cheaper and smaller, thus enabling service providers to enter and exit markets with relative ease. Leasing a transpacific *Internet Protocol* (IP) circuit was much cheaper than leasing a point-to-point *Time Division Multiplexing* (TDM) circuit. For approximately a $100,000 investment (assuming the operator bought the equipment outright), a long-distance bypass operator could compete (though not to scale) with mainstream service providers such as AT&T and Cable and Wireless. Bypass is defined as the provision of telephone service without using the local exchange or toll network of a regulated telephone utility.[1] In this case, the operators were bypassing the *International Message Telephone System* (IMTS), which is largely circuit-switched and not regulated by any one government. Rather it is largely governed by a set of interlocking agreements among international carriers who in turn have correspondent relationships with domestic carriers. Some of those carriers are government owned and are known as *Post, Telegraph, and Telephone* (PTT) organizations. Others are privately owned. Specifically, the VoIP operators were bypassing the international carriers.

The international carriers charged a high rate per minute for international calls. In addition to charging for their own services, they had to pass on what are known as settlements to their interconnecting carriers, in this case U.S.-based carriers who in turn passed the international settlement charges on to the caller. Monies raised via these charges went to governments at the terminating end of the call. By bypassing the IMTS, the VoIP

---

[1]New York State Public Service Commission. "A Glossary of Terms Used by Utilities and Their Regulators." February, 2001. www.dps.state.ny.us/glossary.html.

carriers did not incur charges associated with international settlements and could pass on the savings to their users in the form of lower rates per minute.

In the United States, settlements are governed by the *Federal Communications Commission* (FCC), which directs carriers that have agreements with foreign carriers to report those agreements to the FCC. A more recent agreement governing international settlements is the International Telecommunications Agreement of January 1997.[2]

Two components of technology made these operations possible. First, the VoIP gateways could be purchased for about $200 per port or leased from the gateway vendor for a nominal monthly payment. Second, a transpacific data circuit could be had for a few thousand dollars per month from a data services provider. For some operators, that data circuit was the public Internet. By leasing the VoIP equipment, the operators reduced their sunk costs, a traditional barrier to entry. Sunk costs are defined as expenditures that are at least partially lost once an investment is made (in this case, buying the equipment).[3] A barrier to entry is a constraint that prevents additional sellers from entering a firm's market. In this case, by using low-cost VoIP equipment and leasing data circuits, the service providers could enter the market at a relatively low cost. This scenario also points out a low barrier to exit in that the service provider could leave the market, assuming it became unprofitable, with equal ease.

One case study is that of OpenTel Communications based in San Francisco, California. OpenTel could lease a 128 Kbps leased IP circuit between the United States and Asia for $5,000 per month. Compressing voice at 4.8 Kbps on their Netrix VoIP gateways, OpenTel could get approximately 26 simultaneous conversations on that circuit. OpenTel found that a VoIP solution was faster and less expensive to provision than the leased lines or Frame Relay. At this time, AT&T charged 54 cents per minute for a call from United States to China. Prepaid calling card providers similar to OpenTel charged $.48 per minute.

Once this technology became better known, a race began among operators to enter markets the big three would not service. It became important for operators to establish a *point of presence* (POP) in a given third- or fourth-world nation, connect to a data circuit (often the public Internet or,

---

[2]Kennedy, Charles and Veronica Pastor. *An Introduction to International Telecommunications Law.* Norwood, MA: Artech House, 1996, pg. 123.

[3]"GlobalEdge Glossary." Michigan State University, 2002. http://globaledge.msu.edu/ibrd/glossary.asp?Index = s.

in the case of Africa, Frame Relay via satellite) between that nation and the United States, and commence originating and terminating traffic between the two nations.

A lot of money was at stake for non-U.S. governments trying to block the long-distance bypass operators. Firstly, they were often government-owned monopolies known as PTTs. Long-distance bypass deprived them of valuable revenue. Secondly, the bypass operations deprived the PTTs of settlement payments from the United States.[4] Settlement payments from the States to those PTTs in 1996, for example, totaled about $5.4 billion per year.[5]

Eventually, the long-distance bypass operators achieved legitimacy in their respective markets. The big three carriers soon recognized the advantages of contracting with an operator that has existing facilities in the target countries to handle their traffic between the United States and that nation. This would cause the bypass operators traffic to grow into the many millions of minutes per month of traffic, and it gained them mainstream respectability and the attention of potential buyers. One example is AT&T's March 2000 investment in VoIP carrier Net2Phone.[6]

In the international long-distance bypass industry, VoIP proved to be a disruptive technology in that it had distinct technology deficiencies that were eventually overcome and quantifiable advantages that caused that traffic to prevail over the legacy, mainstream business model. The bypass operators competed directly with mainstream PTTs in international long-distance markets.

International VoIP traffic has continued to grow, increasing from 1.6 billion minutes in 1999 to 5.3 billion in 2000. Although most of this traffic is carried by specialist VoIP carriers, many minutes are originated by traditional *Public Switched Telephone Network* (PSTN) operators who have chosen to outsource some of their international traffic to VoIP operators. On

---

[4]The Accounting Rates System, also known as the International Settlements Policy, sets the accounting rate and the division of tolls. Its origins date to a time when governments owned their respective national monopoly telephone companies. Each carrier and respective government was compensated for originating and terminating a call between their respective countries. The telephone company originating the call billed its customer. That price is known as the collection rate. The price paid to the terminating phone company is known as the accounting rate. Long-distance VoIP bypass carriers avoid paying collection and accounting rates. Source: Kenney and Pastor

[5]Braga, Carlos, Emmanuel Forestier, Charles Kenny, and Peter Smith. "Developing Countries and the Telecommunications Accounting Rates Regime, A Role for the World Bank." World Bank, February 1998. www.worldbank.org/html/fpd/telecoms/pdf/account.pdf.

[6]Boulton, Clint. "AT&T Grabs 37% of Net2Phone." *Internet News*. March 31, 2000. www.internetnews.com/bus-news/article/0,,3_331911,00.html.

the basis of trends in the first half of 2001, international VoIP traffic is likely to reach 10 billion minutes in the current year, equivalent to 6 percent of the world's projected traffic in 2001.[7]

The introduction of the VoIP gateway switch, a component of the softswitch solution and a stepping stone to softswitch architecture, marked the introduction of a destabilizing technology in the international long-distance market. Technology-wise, the VoIP gateway competed directly with the Class 4 switch. Its technology was deficient as compared to the Class 4 in the international market, but given its other advantages some, but not all, operators could overlook those deficiencies. For international long-distance bypass operators, the advantages of being cheaper (in purchase price and *Operations, Administration, Maintenance, and Provisioning* [OAM&P]), smaller (easily deployed anywhere in the world), and more convenient for the bypass operators outweighed any deficiencies (*quality of service* [QoS], small scalability, reliability, and a lack of features). Given a lower cost of long-distance rates, their customers were willing to accept lower QoS. Given the higher margins, the service providers were willing to offer lower QoS. Meanwhile, incumbent service providers lost market share and were forced to cut their prices to compete with the VoIP service providers.

Softswitch technology (a direct evolution from the VoIP gateway) poses the same scenario in the North American converging long-distance market. To apply the comparison of VoIP gateways in the long-distance bypass markets to softswitch in the North American converging markets, it's important to determine if the advantages of softswitch outweigh those of Class 4.

# Softswitch Is More Cost Effective Than Class 4

Softswitch techologies, simply put, are cheaper than circuit switches both in purchase price and in terms of OAM&P. This lowers the barriers to entry to new market entrants, which is critical for competition to enter the local loop market place. The following pages will delve into the cost factors of a voice service provider and how softswitch is the more cost-effective option.

---

[7]TeleGeography, Inc. "TeleGeography 2002—Global Traffic Statistics and Commentary." www.TeleGeography.com, 2001.

# Purchase and OAM&P

Softswitch is cheaper than Class 4 in purchase cost and OAM&P. The net present value models later in this chapter provide concrete comparisons of Class 4 switches versus softswitch in terms of purchase price and OAM&P.

It is estimated that a service provider's largest cost is the ongoing expense of running the network and the switches themselves. Networks must be deployed, maintained, repaired, monitored, and upgraded. The Aberdeen Group has found that competitive carriers spend 60 to 70 percent of their overall expenses on OAM&P. The ability to manage the entire network via a softswitch solution (including adding and altering specific customer services, upgrading capacity, and fixing faults) from a centralized location leads to considerable time and cost savings.[8] Packet-technology-based long-distance carriers maintain that their service provider customers will be able to offer voice services at 20 percent off the traditional rates and they claim the underlying infrastructure can be 40 percent less expensive to operate.[9]

**Figure 11-1**
Projections for softswitch return on investment (ROI) (Source: Frost & Sullivan)

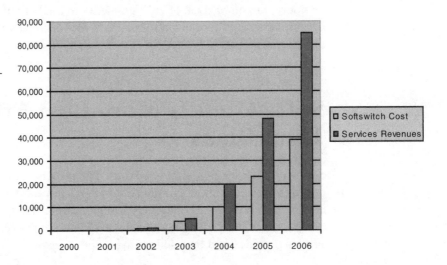

Service Provider Return on Softswitch Investment (In $ Millions)

---

[8]Aberdeen Group, Inc. "Building a Next-Generation Voice Network." February, 2001. www.bitpipe.com.

[9]Krapf, Eric. "VoIP Services Emerging." *Business Communications Review*, January, 2000, p. 17. www.bcr.com/bcrmag/2000/01/p17.asp.

As Figure 11-1 illustrates, in 2001 service providers invested $89 million worldwide on softswitch technology. In the same year, service providers recouped over one-third of that investment in realized service revenue. In the following year, 2002, service providers invested $795 million in softswitch technology. In 2006, service providers will invest over $39 billion in softswitch technology while realizing $85 billion worldwide in revenues for services delivered on that technology.[10]

## Bandwidth Saving

The PSTN transmits conversations at a rate of 64 Kbps. Using a variety of voice *coder/decoders* (codecs), a conversation can be compressed to use less bandwidth. Table 11-1 lists the various *International Telecommunication Union* (ITU) voice codecs and their associated data rates.

The advantages of compressing a conversation is best illustrated by the early long-distance bypass industry. For example, let's say a service provider is providing long-distance service from the United States to Kenya using a 64 Kbps satellite circuit that costs $1,000 per month. Using G.711 uncompressed voice, the service provider can offer only one conversation at a time. Allowing for some degradation of voice quality, the service provider could compress the conversation to 8 Kbps using G.729. Very simply put, the service provider can now get 8 simultaneous conversations over the same 64 Kbps satellite circuit. This has the potential to boost the revenue

**Table 11-1**

*ITU voice codecs and their compression rates*

| Standard | Data rate (Kbps) |
|----------|------------------|
| G.711 | 64 |
| G.721<br>G.723<br>G.726 | 16, 24, 32, 40 |
| G.728 | 16 |
| G.729 | 8 |
| G.723.1 | 5.3, 6.3 |

[10]Dailey, Peter. "The Softswitch—Driving a New Vision of Communication as the Central Element in the Next-Generation Network." A report from Frost & Sullivan, December 2000, pg. 9.

stream for the service provider eightfold and cut the cost of transmission per conversation by a factor of eight. The service provider can determine which compression algorithm offers the best tradeoff in voice quality and revenue.

In the North American market, voice compression in VoIP services has not been a critical issue as the cost of bandwidth, at the time of this writing, has dropped considerably and service providers have been conscious of voice-quality issues when competing with the PSTN. Voice compression is popular for international arbitrage (long-distance bypass) where international bandwidth costs are a serious concern. Although TDM voice can also be compressed, compression has proved popular in VoIP services and is a valid consideration when planning the economics of a softswitched network.

## Lower Barrier to Entry

The overarching issue regarding softswitch being cheaper than Class 4 is that it lowers the cost barrier to enter the converging market. The legacy market has required facilities-based carriers to build much of their network at very fixed high costs. The service providers who actually built long-distance networks could charge very high rates for their circuits as they are a scarce commodity. In the case of long-distance bypass carriers, new operators could enter the market because their cost to do so is much less than if they had to purchase, install, and maintain a Class 4 switch for their operations. The same scenario is slowly materializing in the converging market.

In addition to a lower barrier to entry as regards the acquisition of softswitch, the leasing of IP circuits has also lowered the barrier to entry. As the financial analysis later in this chapter will indicate, the two major cost factors for new market entrants are switches and IP circuits. In most business models, a long-distance service provider will contract with an IP backbone service provider with nationwide coverage (Level 3 and Qwest are good examples) to provide the IP transport for the VoIP service. The long-distance service provider will contract with the IP transport provider for the appropriate bandwidth and a *service level agreement* (SLA).

Another consideration is that mainstream service providers will want to implement softswitch, as it presents a market opportunity for them to lower any costs of expanding their existing operations while lowering costs associated with OAM&P. This further allows them to undercut other mainstream competitors or withstand any price wars initiated by the competition. Adopting softswitch technology can act as an insurance policy against future declines in market prices. By virtue of being able to offer more services than a Class 4 or 5 solution, service providers deploying softswitch can enjoy service differentiation if not additional revenues from the services they offer that nonsoftswitch service providers cannot.

Finally, the creation of new services, from which softswitch-equipped service providers can sell and enjoy higher margins, is another motivating factor for the growth of softswitch sales and will be described later. A look at one's residential or cellular phone bill reveals that the service provider makes most of its money in the margins it makes selling features such as call waiting and voice mail. This is a motivating factor to adopt softswitch.

# Softswitch: A Smaller Footprint

Softswitch is smaller than Class 4 and, as illustrated in Figure 11-2, requires less physical space to deliver the same port density, that is, the equivalent number of phone lines as Class 4. Depending on the configuration, a softswitch may take as little as one-thirteenth of the space required by Class 4 to perform the same service. This is an advantage for softswitch. Due to the smaller footprint, softswitch power consumption and cooling requirements are also less than the legacy switches. Smaller hardware size also translates into lower "real estate" expenses.

One comparison is a softswitch solution from Convergent Networks. Their product can concentrate 36,000 DS0s in one 7-foot rack consuming a

**Figure 11-2**
A comparison of space needed to house 36,000 DS0s in Class 4 and softswitch

Billable space necessary to house 13 racks (36,000 DS0s) for Class 4 – 156 sq.ft.

Space necessary to house 1 rack (36,000 DS0s) for Softswitch - 12 sq. ft

24-inch × 24-inch space, plus space to access the switch, for a billable total of 12 square feet per month. A Class 4 switch in comparison, delivering 36,000 DS0s, requires 13 racks for a total of 156 square feet per month at $35 per square foot of central office space per month. The softswitch in this example offers a 92 percent cost saving over the Class 4 in real estate costs.[11]

The miniaturization of Class 4 components will not soon match the advantages of softswitch in footprint. The smaller footprint of the softswitch translates into lower real estate costs for the service provider. Although this may not be a major concern for an incumbent provider who has built and paid off its central office facilities, it is critical for a new market entrant or an incumbent expanding into new markets. Table 11-2 illustrates the financial advantages in rent or real estate concerns relative to the smaller footprint of softswitch. This advantage is explored further in the net present value models later in this chapter.

Figures 11-3 and 11-4 demonstrate the difference in the central office or collocation space needed to house a Class 4 switch and a softswitch. All the space needed to house the components of either switch generates a monthly rental expense. As set forth in the net present value problems, space for a single rack in the Denver market rents for $1,000 per month. If softswitch requires one rack to do the same function as Class 4, which takes 13 racks, then softswitch costs one-thirteenth as much as Class 4 in monthly rent for this configuration.

The maximum density in a Class 4 switch in one 7-foot rack is 2,688 DS0s. The maximum density of a Convergent Networks ICS2000 softswitch is 108,864 DS0s. In order for the Class 4 to transmit long-distance traffic

**Table 11-2**

A comparison of rental space required between Class 4 and softswitch

|  | Class 4 | Softswitch |
| --- | --- | --- |
| Number of 7-foot racks for 36,000 DS0s | 13 | 1 |
| Rental cost per month per rack | $1,100 | $1,100 |
| Total per month | $14,300 | $1,100 |
| Total per year per location | $171,600 | $13,200 |

[11]Convergent Networks. "A Business Case for Next-Generation Switching: Revolutionizing Carrier Revenues and Returns." pg. 3. www.convergentnet.com.

**Figure 11-3**
The City of Denver's Legacy Nortel TDM PBX (Source: William Herwig, Cisco Systems)

**Figure 11-4**
The City of Denver's new Cisco IP PBX solution. Note the size comparison with TDM PBX in Figure 11-3 (Source: William Herwig, Cisco Systems).

via an IP network, the Class 4 would require VoIP media gateways requiring even more rack space.

The smallest configuration for DMS-250 (Nortel's Class 4) is 480 DS0s (20 T1s). For many locations and service providers, this presents a potentially expensive overabundance of capacity. Many media gateways scale by the T1 or E1 card, enabling a greater flexibility in scaling down to capture certain markets and to take advantage of distributed architecture. If a service provider purchases and installs only what they need to get started in a given market and add capacity only as it is absolutely necessary, then capital is not invested in equipment that cannot immediately generate revenue.

The capability of a softswitch to scale *down* is an advantage for many carriers. The smallest configuration for the DMS-250 Class 4 switch, for example, is 20 T1s. Many media gateways scale *down* to one T1 or less. A low-density media gateway could be installed in an outlying city, for example, and negate the need to backhaul long-distance traffic from that outlying city to a larger city where the Class 4 switch is collocated with a Class 5 switch in a central office. This eliminates expensive "hairpins" in traffic. Given the distributed architecture of the softswitch solution, the *media gateway controller* (MGC) and signaling gateway can be located elsewhere. This flexibility can also allow new operators to enter markets more easily. Chapter 6, "Softswitch: More Scalable Than CLASS 4 or 5," covered scalability advantages of softswitch solutions.

# Softswitch Advantages in Power Draw

With fewer assets to power, softswitch uses less power. This translates into financial savings the service provider cannot ignore. A Class 4 configuration similar to that in Table 11-3 might draw 600 amps DC, as opposed to a comparable softswitch configuration that draws 24 amps DC. This translates into softswitch using only four percent of the power of Class 4.[12] These sav-

---

[12]Interview with Nathan Stratton, CTO, Exario (www.exario.com), October 28, 2001.

**Table 11-3**

Comparison of
Class 4 and
softswitch power
draw and cost of
power

| | Class 4 | Softswitch |
|---|---|---|
| Number of 7-foot racks for 36,000 DS0s | 13 | 1 |
| Cost per month per rack for power at 20 amps at $20/amp | $400 | $400 |
| Total per month | $5,200 | $400 |
| Total per year per location | $62,400 | $4,800 |

ings will be explored further in the net present value models at the end of this chapter.

## Advantages of Distributed Architecture

Given a shift in the paradigm regarding distributed versus centralized architecture, it is possible that the domestic U.S. market might go the way of the long-distance bypass market. As described previously, long-distance bypass carriers were enabled by the introduction of smaller gateways to roll out service in diverse markets. They took market share from large, mainstream, long-distance providers. Let's assume service providers who were not traditionally voice service providers, that is, cable TV operators, *Internet service providers* (ISPs), power companies, or *digital subscriber line* (DSL) providers, wanted to originate and terminate long-distance traffic for their existing customers. Following the long-distance bypass model, they could install media gateways in a number of POPs in Tier 1 and 2 cities, and they could potentially take market share from the legacy long-distance Class 4 carriers. This scenario points to the displacement of Class 4 by softswitch, a disruptive technology.

Figure 11-5 details how the components of softswitch interoperate. Each element can be located remotely from any of the others. As Figure 11-6 illustrates, distributed architecture enables a service provider to install only media gateways in each area serviced, as opposed to a Class 4 that requires an entire switch be installed in each area serviced. Given the difference in footprint and the accompanying cost, as discussed earlier, this can lead to substantial savings for the service provider.

**Figure 11-5**
The components of
softswitch are
distributed.

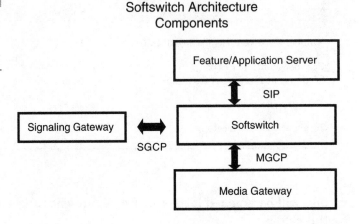

**Figure 11-6**
A distributed
architecture enables
the dispersal of
softswitch solution
components that
can lower real estate
costs.

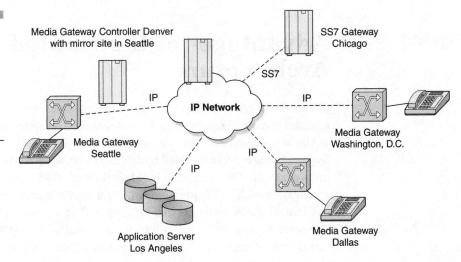

# Economic and Regulatory Issues Concerning Softswitch

In its April 10, 1998 Report to Congress, the FCC determined that phone-to-phone IP telephony is an enhanced service and is not a telecommunications service. The important distinction here is that telecommunications service providers are liable for access charges to local service providers both at the originating and terminating ends of a long-distance call. A telecommunications service provider must also pay into the Universal Service

Fund. Long-distance providers using VoIP (and by inference, softswitch) avoid paying access and Universal Service fees. Given thin margins on domestic long distance, this poses a significant advantage for phone-to-phone IP telephony service providers.[13]

The possibility that the FCC may rule differently in the future cannot be discounted. Having to pay access fees to local carriers to originate and terminate a call coupled with having to pay into the Universal Service Fund would pose a significant financial risk to the business plan of a softswitch-equipped, VoIP, long-distance service provider. Just as international long-distance bypass providers used VoIP to bypass international accounting rates and make themselves more competitive than circuit-switched carriers, softswitch-equipped VoIP carriers can make themselves more competitive in the domestic market by bypassing access charges and avoiding paying into the Universal Service Fund. The service provision model set forth in this chapter is strongly affected by the possibility of the FCC reversing itself on phone-to-phone IP telephony.

Access fees in North American markets run from about $.01 per minute for origination and termination fees to upwards of $.05 per minute in some rural areas. That is, a call originating in Chicago, for example, would generate an origination fee of $.01 per minute. If the call terminated in Plentywood, Montana, it may generate a $.05 per minute termination fee. This call would generate a total of $.06 per minute in access fees. If the carrier can only charge $.10 per minute, it will reap only $.04 per minute for this call after paying access fees to the generating and terminating local phone service providers.

Table 11-4 illustrates the impact on profits and losses for a long-distance service provider that must pay access fees. The impact of the access fees on the net present value of VoIP carriers who are exempt from access fees and non-VoIP carriers is addressed later in this chapter where a service provider generates 25 percent more revenue by virtue of not paying access fees to other carriers. It is possible that the FCC at some point could reverse this ruling and make VoIP carriers pay access fees.

## Net Present Value of Softswitch

The net present value is an engineering economics term for determining when the benefit of investing in a new technology outweighs the cost of

---

[13]"Federal Communications Commission Report to Congress," April 10, 1998, paragraphs 88–93.

**Table 11-4**

The impact of access fees on long-distance revenues

| Retail price (cents per minute) | Origination fee | Termination fee | Revenue after access fees | Access fees as a percentage of retail price |
|---|---|---|---|---|
| 5 | 1 | 1 | 3 | 40 |
| 5 | 1 | 5 | −1 | 120 |
| 5 | 5 | 5 | −5 | 200 |
| 10 | 1 | 1 | 8 | 20 |

doing so. Consider a service provider who is considering entering the converging market. In order to affect an accurate apples-to-apples comparison in the following problems, it is assumed that the Class-4-equipped carrier uses IP for long-distance transport and is equipped with media gateways to perform the TDM-IP-TDM translations. For the softswitch solution they would have to invest in

- Media gateways
- Signaling gateways
- Application server
- Media server
- Softswitch

For the Class 4 solution for a converging market project, they would have to invest in

- Class 4 switch
- Media gateways to handle the TDM to IP translation

How does the firm determine which is better from a dollar and cents point of view? First, it is important to lay out the economic components in this decision-making process.

The planners need to work two different calculations. One is the net present value of the softswitch option and the other is the net present value of the Class 4 option. The elements of the equation are the investment outlay (cost) of $P < O$, which generates a series of future benefits $(B > O)$ and costs $(C < O)$ for $j = 0, 1, 2 \ldots, N$ time periods. For the sake of brevity and paucity of year-to-year cost and benefit information, we will assume the annual costs and benefits remain unchanged year to year. Accordingly the net present value formula, $NPV = (B_N + C_N)(P/A, i, N) + P$, is used.

# Considerations for the Net Present Value Models

These models will compare three different long-distance providers, one small long-distance operator (480 DS0s), a mid-sized long-distance operator (4,032 DS0s), and a major long-distance carrier (36,000 DS0s). For brevity, the number of DS0s studied is per POP (a node on their network) and does not address the service provider's entire network. The pricing of the Class 4 switching solutions is based on testimony before the FCC where small configurations were priced at $500,000 for the basic chassis and $100 per DS0. Large configurations (100,000 DS0s) are priced at $1,000,000 per chassis plus $100 per DS0.[14] For the Class 4 switch, the Nortel DMS-250 is used strictly as an example of a Class 4 switch. The pricing of the softswitch solutions was provided by the softswitch vendors and is considered "street pricing."

The tables (in the following pages) detailing the comparison of acquisition costs list all the factors necessary to complete an installation of the respective switches. The parts contained are the system itself, an estimate of installation costs based on a per diem charge (often $1,500 per day per installer), and sales tax (state and local). Following the acquisition costs ($P = C_0$), the operating expenses for the life of the project ($N = 2$ for a period of two years) are detailed. Sources for further costing information are as follows:

- **Installation expenses at $1,500/day**   Nuera
- **Acquisition costs**   Respective softswitch vendors for softswitch costs
- **Class 4 acquisition costs**   Dr. Robert Mercer from testimony before the FCC
- **Operating expenses**   WorldCom and Level3
- **Labor costs**   ICG

---

[14]Interview with Dr. Robert Mercer, Hatfield Consulting, November 1, 2001. Nortel Networks Sales did not respond to repeated inquiries.

Annual operating costs ($C_N$) include rental costs, labor, data circuits, and leasing costs. Real estate refers to the rental costs of locating a seven-foot equipment rack in a collocation space. Both WorldCom and Level3 charge about $1,000 per month per rack in their Denver collocation spaces. Power is included in that charge. Labor is differentiated in that a Class 4 will usually require an engineer where the industry salary is about $85,000 per year. Softswitches can be maintained by a technician, most of whom are compensated at a maximum of $70,000 per year.

Maintenance contracts throughout the telecommunications industry are usually ranked into three categories from the least expensive—"Bronze" with the least amount of service (24/7 phone support only) at about 10 percent of the purchase price per year, up to "Gold" service (on site service with "spares in the air") that hits on 30 percent of the purchase price per year. In between is "Silver" service that earns 20 percent per year and is a step up in services from "Bronze." For this study, 10 percent is used. Data circuit pricing is provided by WorldCom for service in Denver.

*Opportunity cost of capital* (OCC), or $i$, is set at 7 percent, 10 percent, and 12 percent, as it assumes at the lower end that entrepreneurs cannot achieve much more with a certificate of deposit and at the higher end venture capitalists in the 2001 stock market cannot achieve any greater return on their money. The study sets N, the life of the operation or the operational life of the Class 4 switch and softswitch, at two years. This was arrived at because, given the pace of change in the datacom and telecom industry, the softswitch platforms will be outdated by that time. Also, software versions contained on the Class 4 could very well be out of date in two years as well.

Two years will appear to be a short operational life for a telecommunications project, but given the type of technology involved and the pace of change in this industry, two years is appropriate. This may contrast sharply with legacy telecommunications depreciation schedules that run into decades and highlights the seismic change occurring in this industry. The first consideration is that much of the technology contained in both technologies is dependent upon semiconductors.

Moore's Law states that computing power (the power of semiconductors) doubles every 18 to 24 months, and the price of that computing power has decreased 54.1 percent each year from 1985 to 1996. At the same time, the Gross Domestic Product deflator has increased by 2.6 percent per year. Communications equipment, in this case both Class 4 and softswitch, is based largely on semiconductors. Hence, Moore's Law applies to both softswitch and Class 4, but more so to softswitch, given its client/server architecture (a newer platform than Class 4 main frame).

Moore's Law, as it would apply to softswitch applications, has been exceeded by the growth in DS0 density in media gateways and the expansion in the number of *Busy Hour Call Attemps* (BHCAs). The count of DS0s per 7-foot rack has grown from 1,176 for a Netrix media gateway in 1998 to 108,864 for a Convergent Networks media gateway in 2001, marking a growth in capacity of 30 times per year. In addition, Dale W. Jorgenson of *American Economic Review* states that software investment is growing rapidly and is now much more important than investments in computer hardware. Given that the main cost of a softswitch solution is software, this is very true of softswitch. If these trends continue, it is wise to use two years as a life cycle for these projects.[15]

Annual benefit $(B_N)$ is calculated as the respective number of DS0s busy 8 hours per business day $(8 \times 60 = 480$ minutes per business day) at 100 percent with 254 business days per year at $.03 per minute. Three cents per minute is chosen to reflect either a wholesale operation where a reseller has to remain competitive in the North American market, the customer is a large corporate user of long distance, or the market is very competitive.

It should be noted that for the purposes of this net present value study the Class 4 and the softswitch deliver the same service with the same features at the same level of quality. The end users cannot discern if their call is routed via a softswitch solution or a Class 4.

Net present value is either a positive or negative value. In the case of this analysis, the switch with the highest number is the most positive or best choice in delivering value for the investment. A high net present value means more wealth and higher returns to shareholders. It also assumes no constraint on financial capital. Thus, a firm has cash available to make transactions. No loans are required to finance transactions.

Table 11-5 lists purchase installation costs for both Class 4 and softswitch. Installation figures are included because when a company leases equipment the cost of installation can be included in the lease, making it a business expense.

Table 11-6 details the operating expenses of Class 4 and softswitch, which for income tax purposes translates into business expenses. Table 11-7 outlines the gross revenue for 480 DS0s, and Table 11-8 illustrates the cost structures when leasing Class 4 and softswitch for 480 DS0s.

---

[15]Jorgenson, Dale W. "Information Technology and the U.S. Economy." *American Economic Review*, Vol. 91, 2001.

**Table 11-5**

Purchase costs of Class 4 and softswitch for 480 DS0s

| Components of cost | Class 4 DMS-250 | Softswitch NuTANDEM |
|---|---|---|
| System | $638,000 | $200,000 |
| Installation | $7,500 | $7,500 |
| Total (P = Co) | $645,500 | $207,500 |

**Table 11-6**

Class 4 and softswitch business expenses for the project's lifetime at 480 DS0s

| Component of Cost | Class 4 DMS-250 | Softswitch NuTANDEM |
|---|---|---|
| Collocation rent (annual) | $12,000 | $12,000 |
| Labor | $85,000 (engineer) | $70,000 (technician) |
| Maintenance contract | $55,000 | $20,000 |
| IP circuits | $120,000 | $120,000 |
| Total | $272,000 | $222,000 |
| For life (j) of 2 years | $544,000 | $444,000 |

**Table 11-7**

Gross revenue for 480 DS0s

| DS0s | Minutes/day | Price/minute | Days (2 yrs) | Total benefit |
|---|---|---|---|---|
| 480 | 480 | .03 | 508 | $3,511,296 |

**Table 11-8**

Cost structures when leasing Class 4 and softswitch for 480 DS0s

| Component of cost | Class 4 DMS-250 | Softswitch NuTANDEM |
|---|---|---|
| Collocation rent (annual) | $12,000 | $12,000 |
| Labor | $85,000 (engineer) | $70,000 (technician) |
| Maintenance contract | $63,800 | $20,000 |
| IP circuits | $120,000 | $120,000 |
| Lease cost | $45,190 | $14525 |
| Total | $325,950 | $236,525 |
| For life (j) of 2 years | $651,900 | $473,050 |

Table 11-9 details the benefits and costs of Class 4 and softswitch for a configuration with 480 DS0s. Note that leasing is included as a business expense. The net present value is determined as follows: $NPV = (B_N + C_N)(P/A, i, N) + P$ where the opportunity cost of capital is 12 percent. Table 11-10 demonstrates how to determine NPV when leasing for Class 4 and softswitch. Table 11-11 shows the net present value when leasing at 12, 10, and 7 percent OCC for leasing 480 DS0s. Table 11-12 outlines the total costs and benefits when purchasing Class 4 or softswitch for 480 DS0s. Table 11-13 shows the salvage values for 480 Class 4 and softswitch DS0s.

**Table 11-9**

Total benefits and costs of Class 4 and softswitch for 480 DS0s when leasing

| Income tax component | Class 4 | Softswitch |
|---|---|---|
| Gross income | $3,511,296 | $3,511,296 |
| Business expenses (includes leasing costs) | −$651,900 | −$473,050 |
| Taxable income | $2,889,246 | $3,038,246 |
| Income tax as figured from federal income tax rates for corporation, 1994, Instructions for forms 1120 and 1120-A | $113,900 + 34% in excess over $335,000 = $113,900 + ($2,889,246 − 335,900)(.34) = $982,038 | $113,900 + 34% in excess over $335,000 = $113,900 + ($3,038,246 − 335,900)(.34) = $1,032,697 |
| Profit after taxes and expenses | $1,887,507 | $2,005,549 |

**Table 11-10**

Determining NPV when leasing

| Class 4 | Softswitch |
|---|---|
| $NPV = (\$2,889,246 - \$982,038)(1.6901)$ = $3,190,076 | $NPV = (\$3,038,246 - \$1,032,697)(1.6901) = \$3,389,578$ |

**Table 11-11**

Net present value when leasing at 12%, 10%, and 7% OCC for leasing 480 DS0s

| Class 4 | Softswitch |
|---|---|
| $3,190,076 where N = 2, OCC = 12 | $3,389,578 where N = 2, OCC = 12 |
| $3,275,768 where N = 2, OCC = 10 | $3,480,630 where N = 2, OCC = 10 |
| $3,412,613 where N = 2, OCC = 7 | $3,626,032 where N = 2, OCC = 7 |

**Table 11-12**

Total costs and benefits when purchasing Class 4 or softswitch for 480 DSOs

| Income tax component | Class 4 | Softswitch |
|---|---|---|
| Gross income | $3,511,296 | $3,511,296 |
| Business expenses | −$561,600 | −$444,000 |
| Taxable income before depreciation and/or capital gains | $2,949,696 | $3,038,246 |
| Depreciation | −$496,236 | −$155,560 |
| Salvage value at year two | $278,193 | $94,755 |
| Book value at year two | $210,540 | $44,440 |
| Difference salvage value versus book value | $67,653 taxable as capital gains at 20% = $13,530 | $50,315 taxable as capital gains at 20% = $10,063 |
| Tax on capital gains | $13,530 | $10,063 |
| Income tax as figured from federal income tax rates for corporation, 1994, instructions for forms 1120 and 1120-A | $113,900 + 34% in excess of over $335,000 = $113,900 + ($2,949,696 − 335,900)(.34) = $1,002,590 + capital gains $13,530 = $1,016,120 | $113,900 + 34% in excess over $335,000 = $113,900 + ($3,038,246 − 335,900)(.34) = $1,032,698 + capital gains at $10,063 = $1,042,760 |
| Profit after taxes and expenses | $1,933,575 | $1,995,485 |

**Table 11-13**

Salvage values for 480 Class 4 and softswitch DSOs

| Salvage value class 4 (P-S)(A/P, I, N) + Si | Salvage value softswitch |
|---|---|
| ($550,000 − $100,000)(.5917) + $100,000(.12) = $278,265 | ($200,000 − $50,000)(.5917) + $50,000(.12) = $94,755 |

Table 11-14 lists the net present values when purchasing 480 DSOs. Note in Table 11-12 the approximately 17 percent greater value of softswitch over Class 4 in net present value.

Tables 11-11 and 11-14 detail the sensitivity analysis for differences in the OCC of the Class 4 versus softswitch project. In each instance, softswitch still has a more positive value than Class 4, which determines that softswitch is the better value in this analysis.

**Table 11-14**

Net present value when purchasing 480 DSOs

| Class 4 NPV with OCC at 12%, 10%, and 7% | Softswitch NPV with OCC at 12%, 10%, and 7% |
|---|---|
| NPV = ($1,933,575)(1.6901) − $638,000 = $2,631,675 | NPV = ($1,995,485)(1.6901) − $200,000 = $3,172,569 |
| NPV = ($1,933,575)(1.7355) − $638,000 = $2,717,719 | NPV = ($1,995,485)(1.7355) − $200,000 = $3,263,164 |
| NPV = ($1,933,575)(1.8080) − $638,000 = $2,857903 | PV = ($1,995,485)(1.8080) − $200,000 N= $3,407,836 |

Table 11-15 provides depreciation information on Class 4 and softswitch for a two-year project. Table 11-16 compares acquisition costs for used Class 4 versus new softswitch. Table 11-17 shows the total benefits and cost when leasing used Class 4 versus softswitch. In Table 11-18, Softswitch delivers a slightly higher value at this scale. Table 11-19 illustrates income tax implications with depreciation and the salvage value for used Class 4 and softswitch. The net present value when purchasing used Class 4 and new softswitch is shown in Table 11-20.

Note that when buying a used Class 4 at $100,000 or half the purchase price of a softswitch solution, Class 4 offers only a slightly higher net pre-

**Table 11-15**

Depreciation on Class 4 and softswitch for a two-year project

| Depreciation of Class 4 (GDS method) over 2 years (Year 1 + Year 2) | Depreciation of softswitch (GDS method) over 2 years |
|---|---|
| 550,000(.3333) + $550,000(.4445) = $427,790 | $200,000(.3333) + $200,000(.4445) = $155,560 |

**Table 11-16**

A comparison of acquisition costs for used Class 4 versus new softswitch

| Components of Cost | Used DMS-250 | Softswitch NuTANDEM |
|---|---|---|
| System | $100,000 | $200,000 |
| Installation | $7,500 | $7,500 |
| Total (P = Co) | $107,500 | $207,500 |

**Table 11-17**

Total benefits and cost when leasing used Class 4 versus softswitch

| Income tax component | Class 4 | Softswitch |
|---|---|---|
| Gross income | $3,511,296 | $3,511,296 |
| Business expenses (includes leasing costs) | −$622,050 | −$473,050 |
| Taxable income | $2,889,246 | $3,038,246 |
| Income tax as figured from federal income tax rates for corporation, 1994, instruction for forms 1120 and 1120-A | $113,900 + 34% in excess over $335,000 = $113,900 + ($2,889,246 − 335,900) (.34) = $982,343 | $113,900 + 34% in excess over $335,000 = $113,900 + ($3,038,246 − 335,900)(.34) = $1,032,697 |
| Profit after taxes and expenses | $1,906,903 | $2,005,549 |

**Table 11-18**

Net present value when leasing used Class 4 and new softswitch.

| Class 4 | Softswitch |
|---|---|
| NPV = ($1,906,903)(1.6901) = $3,222,856 | NPV = ($2,005,549)(1.6901) = $3,389,578 |
| NPV = (1,906,903)(1,7355) = $3,309,430 | NPV = ($2,005,5490(1.7355) = $3,480,302 |
| NPV = ($1,906,903)(1.8080) = $3,447,680 | NPV = ($2,005,549)(1.8080) = $3,626,032 |

**Table 11-19**

Income tax implications with depreciation and salvage value for used Class 4 and softswitch

| Income tax component | Class 4 | Softswitch |
|---|---|---|
| Gross income | $3,511,296 | $3,511,296 |
| Business expenses | −$544,000 | −$444,000 |
| Taxable income before depreciation and/or capital gains | $2,967,296 | $3,038,246 |
| Depreciation | −$77780 | −$155,560 |
| Salvage value | $54,453 | $94,755 |
| Book value | $22,220 | $44,440 |
| Difference salvage value vs. book value | $32,233 taxable at 20% capital gains | $50,315 taxable at 20% |
| Taxable on capital gains | $6,446 | $10,063 |

**Table 11-19 cont.**

Income tax implications with depreciation and salvage value for used Class 4 and softswitch

| Income tax component | Class 4 | Softswitch |
|---|---|---|
| Income tax as figured from federal income tax rates for corporation, 1994, instructions for forms 1120 and 1120-A | $113,900 + 34% in excess over $335,000 = $113,900 + ($2,999,529 − 335,900)(.34) = $1,019,533 + $6,446 = $1,025,979 | $113,900 + 34% in excess over $335,000 = $113,900 + ($3,138,876 − 335,900)(.34) = $1,066,911 + $10,063 = $1,076,974 |
| Profit after taxes and expenses | $1,941,317 | $1,961,272 |

**Table 11-20**

Net present value when purchasing used Class 4 and new softswitch

| Class 4 net present value with OCC at 12%, 10%, and 7% | Softswitch NPV with OCC at 12%, 10%, and 7% |
|---|---|
| NPV = ($1,941,317)(1.6901) − $100,000 = $3,181,020 | NPV = ($1,961,272)(1.6901) − $200,000 = $3,114,745 |
| NPV = ($1,941,317)(1.7355) − $100,000 = $3,269,155 | NPV = ($1,961,272)(1.7355) − $200,000 = $3,203,787 |
| NPV = ($1,941,317)(1.8080) − $100,000 = $3,409,901 | NPV = ($1,961,272)(1.8080) − $200,000 = $3,345,997 |

sent value than softswitch. When leasing used Class 4 versus new softswitch, softswitch delivers slightly better value.

## Net Present Value: Midsized Long-Distance Service Provider

This net present value model will compare Class 4 with the softswitch product from Convergent Networks, the ISC2000, which is an example of a more densely populated (the media gateway, that is) softswitch solution. Table 11-21 shows the acquisition costs of Class 4 and softswitch for 4,032 DS0s, while Table 11-22 provides the cost of Class 4 and softswitch for 4,032 DS0s. Table 11-23 illustrates the gross revenue for 4,032 DS0s.

**Table 11-21**

Acquisition costs of
Class 4 and
softswitch for
4,032 DS0s

| Components of cost | Class 4 DMS-250 | Softswitch Convergent Networks ISC2000 |
|---|---|---|
| System | $1,085,000 | $225,000 |
| Installation | $15,000 | $9,000 |
| Total (P = Co) | $1,100,000 | $234,000 |

**Table 11-22**

Cost of Class 4 and
softswitch for
4,032 DS0s

| Components of cost | Class 4 DMS-250 | Softswitch Convergent Networks ISC2000 |
|---|---|---|
| Collocation rent (annual) | $48,000 | $12,000 |
| Labor | $85,000 (engineer) | $70,000 (technician) |
| Maintenance contract | $108,500 | $22,500 |
| IP circuits | $1,080,000 | $1,080,000 |
| Total | $1,321,500 | $1,239,100 |
| For life (j) of 2 years | $2,643,000 | $2,478,200 |

**Table 11-23**

Gross revenue for
4,032 DS0s

| DS0s | Minutes/day | Price/minute | Days (2 yrs) | Total benefit |
|---|---|---|---|---|
| 4,032 | 480 | .03 | 508 | $29,494,886 |

# Leasing Class 4 Versus Softswitch

Table 11-24 details the cost structure with leasing Class 4 and softswitch for 4,032 DS0s. The benefits and costs when leasing Class 4 and softswitch for 4,032 DS0s are shown in Table 11-25.

Table 11-26 compares net present values for Class 4 and softswitch when leasing 4,032 DS0s. Softswitch holds a slight (approximately 2 percent) edge over Class 4.

**Table 11-24**

Cost structure with leasing Class 4 and softswitch for 4,032 DS0s

| Component of cost | Class 4 DMS-250 | Softswitch Convergent Networks ISC2000 |
|---|---|---|
| Collocation rent (annual) | $48,000 | $12,000 |
| Labor | $85,000 (engineer) | $70,000 (technician) |
| Maintenance contract | $108,500 | $22,500 |
| Leasing | $75,950 | $15,750 |
| IP circuits | $1,080,000 | $1,080,000 |
| Total | $1,397,450 | $1,200,250 |
| For life (j) of 2 years | $2,794,900 | $2,400,500 |

**Table 11-25**

Benefits and costs when leasing Class 4 and softswitch for 4,032 DS0s

| Income tax component | Class 4 | Softswitch |
|---|---|---|
| Gross income | $29,494,886 | $29,494,886 |
| Business expenses | ($2,794,900) | ($2,400,500) |
| Taxable income before depreciation and/or capital gains | $26,699,986 | $27,094,386 |
| Income tax | $26,699,986(.35) = $9,344,995 | $27,094,386(.35) = $9,483,035 |
| Profit after taxes and expenses | $17,354,991 | $17,611,351 |

**Table 11-26**

Net present value when leasing Class 4 and softswitch for 4,032 DS0s

| Class 4 NPV with OCC at 12%, 10% and 7% | Softswitch NPV with OCC at 12%, 10% and 7% |
|---|---|
| NPV = ($17,354,991)(1.6901) = $29,331,670 | NPV = ($17,611,351)(1.6901) = $29,764,944 |
| NPV = ($17,354,991)(1.7355) = $30,119,586 | NPV = ($17,611,351)(1.7355) = $30,564,499 |
| NPV = ($17,354,991)(1.8080) = $31,377,823 | NPV = ($17,611,351)(1.8080) = $31,841,322 |

# Buying Class 4 or Softswitch 4,032 DS0s

The previous case studied the Net Present Value of leasing a Class 4 and softswitch. The following case studies purchasing these platforms. Leasing allows an operator to enter a market without the upfront cost of purchasing a switch. Another difference between leasing and purchasing a switch is that a switch that has been purchased as opposted to leased retains some salvage value.

Table 11-27 outlines the benefits and costs when purchasing Class 4 or softswitch for 4,032 DS0s. The net present value of Class 4 and softswitch when purchasing 4,032 DS0s is the subject of Table 11-28. As with Table 11-26, Table 11-28 reveals a 3 percent advantage of softswitch over Class 4 when purchasing both platforms. Table 11-29 illustrates the salvage value for Class 4 and softswitch for 4,032 DS0s.

**Table 11-27**

Benefits and costs when purchasing Class 4 or softswitch for 4,032 DS0s

| Income tax component | Class 4 | Softswitch |
|---|---|---|
| Gross income | $29,494,886 | $29,494,886 |
| Business expenses | ($2,643,000) | ($2,478,200) |
| Taxable income before depreciation and/or capital gains | $26,851,886 | $27,016,686 |
| Depreciation | ($843,912) | ($175,005) |
| Salvage value | $473,102 | $189,976 |
| Book value | $358,050 | $74,250 |
| Difference salvage value vs. book value | ($115,052) | ($115,726) |
| Tax on capital gains (20% of above) | $23,010 | $23,145 |
| Income tax | $26,851,886(.35) = $9,398,160 | $9,455,840 |
| Profit after taxes and expenses | $17,453,726 | $17,537,701 |

**Table 11-28**

Net present value of Class 4 and softswitch when purchasing 4,032 DS0s

| Class 4 NPV with OCC at 12%, 10%, and 7% | Softswitch NPV with OCC at 12%, 10% and 7% |
|---|---|
| NPV = ($17,453,726)(1.7355) − $1,085,000 = $29,205,941 | NPV = ($17,537,701)(1.7355) − $225,000 = $30,211,680 |
| NPV = ($17,453,726)(1.6901) − $1,085,000 = $28,413,542 | NPV = ($17,537,701)(1.6901) − $225,000 = $29,415,468 |
| NPV = ($17,453,726)(1.8080) − $1,085,000 = $30,471,337 | NPV = ($17,537,701)(1.8080) − $225,000 = $31,483,163 |

**Table 11-29**

Salvage value for Class 4 and softswitch for 4,032 DS0s

| Salvage value Class 4 (P-S)(A/P, I, N) + Si | Salvage value softswitch |
|---|---|
| ($1,085,000 − $358,050)(.5917) + $358,050(.12) = $473,102 | ($225,000 − $74,250)(.5917) + $74,250(.12) = 189,976 |

# Buying Class 4 or Softswitch with Class 4 at $50 per DS0

The following net present value model is provided to demonstrate net present values when Class 4 vendors drastically lower their prices. This model reflects a sale price of $50 per DS0 for the Class 4 switch. In order to transmit voice over an IP network, the Class 4 switch also requires a VoIP gateway, as detailed in Table 11-30. Table 11-31 shows the operating expenses for Class 4 at $50 per DS0 versus softswitch. Table 11-32 outlines the costs

**Table 11-30**

Cost structure for Class 4 at $50 per DS0 plus a necessary VoIP gateway versus softswitch

| Components of cost | Class 4—DMS-250 | Softswitch Convergent Networks ISC2000 |
|---|---|---|
| System | $50/DS0 × 4,032 + VoIP gateway at $182,000 = $383,600 | $225,000 |
| Installation | $15,000 | $9,000 |
| Total (P = Co) | $398,000 | $234,000 |

**Table 11-31**

Operating expenses for Class 4 at $50 per DS0 versus softswitch

| Components of cost | Class 4—DMS-250 | Softswitch Convergent Networks ISC2000 |
|---|---|---|
| Collocation rent (annual) | $48,000 | $12,000 |
| Labor | $85,000 (engineer) | $70,000 (technician) |
| Maintenance contract | $38,360 | $22,500 |
| IP circuits | $1,080,000 | $1,080,000 |
| Total | $1,251,360 | $1,239,100 |
| For life (j) of 2 years | $2,502,720 | $2,478,200 |

**Table 11-32**

Costs and benefits of buying Class 4 at $50 per DS0 plus a VoIP gateway

| Income tax component | Class 4 | Softswitch |
|---|---|---|
| Gross income | $29,494,886 | $29,494,886 |
| Business expenses | ($2,502,720) | ($2,478,200) |
| Taxable income before depreciation and/or capital gains | $26,992,166 | $27,016,686 |
| Salvage value | $176,264 | $189,976 |
| Book value | $115,080 | $74,250 |
| Difference salvage value versus book value | ($61,184) | ($115,726) |
| Tax on capital gains (20% of above) | $12,236 | $23,145 |
| Income tax | $26,992,166(.35) = $9,447,258 | $9,455,840 |
| Profit after taxes and expenses | $17,532,672 | $17,537,701 |

and benefits of buying Class 4 at $50 per DS0 plus a VoIP gateway. The net present value of Class 4 at $50 per DS0 with a VoIP gateway versus softswitch is covered in Table 11-33.

It is interesting to note that even with the Class 4 at $50 per DS0, it does not exceed the softswitch in net present value. However, the difference between the two platforms is inconsequential.

| **Table 11-33** | Class 4 NPV with OCC at 12%, 10%, and 7% | Softswitch NPV with OCC at 12%, 10%, and 7% |
|---|---|---|
| NPV Class 4 at $50 per DS0 plus a VoIP gateway versus softswitch | NPV = ($17,532,672)(1.6901) − $383,600 = $29,248,368 | NPV = ($17,537,701)(1.6901) − $225,000 = $29,415,468 |
| | NPV = ($17,532,672)(1.7355) − $383,600 = $30,044,352 | NPV = ($17,537,701)(1.7355) − $225,000 = $30,211,680 |
| | NPV = ($17,532,672)(1.8080) − $383,600 = $31,315,470 | NPV = ($17,537,701)(1.8080) − $225,000 = $31,483,163 |

# Net Present Value When Softswitch Generates Greater Revenue

According to Laura Thompson, Vice President of Marketing and Channels at softswitch vendor Sylantro (www.sylantro.com), a softswitch-enabled service provider could enjoy up to 25 percent greater revenue by virtue of offering high-margin calling features that a Class 4 or 5 service provider cannot offer. The following net present value model will determine if a higher net present value is achieved where the softswitch-enabled service provider is enjoying revenues greater than that of the Class-4-equipped service provider. Table 11-34 outlines the costs and benefits when softswitch generates 25 percent more revenue than Class 4, while Table 11-35 shows the NPV when softswitch generates 25 percent more revenue than Class 4.

The previous tables denote the net present value when softswitch generates 25 percent more revenue than Class 4. In this model, the net present values for softswitch are over 30 percent more for softswitch than for Class 4. This is a significant difference that could sway decision makers to choose softswitch over Class 4.

The previous net present value model also illustrates the advantages a service provider could enjoy when not paying access fees or paying into the Universal Service Fund. For this model, it is assumed that the Class 4 solution is not a phone-to-phone IP service and it is not considered an enhanced service provider; it must pay access fees and pay into the Universal Service Fund. The softswitch-enabled service provider is operating as a phone-to-phone IP service provider and is exempt from paying origination and termination fees to local service providers. The gross income figure is the

**Table 11-34**

Costs and benefits when softswitch generates 25% more revenue than Class 4

| Income tax component | Class 4 | Softswitch |
|---|---|---|
| Gross income | $29,494,886 | $36,868,608 |
| Business expenses | ($2,643,000) | ($2,478,200) |
| Taxable income before depreciation and/or capital gains | $26,851,886 | $34,390,408 |
| Depreciation | ($843,912) | ($175,005) |
| Salvage value | $473,102 | $189,976 |
| Book value | $358,050 | $74,250 |
| Difference salvage value versus book value | ($115,052) | ($115,726) |
| Tax on capital gains (20% of above) | $23,010 | $23,145 |
| Income tax | $26,851,886(.35) = $9,398,160 | $12,036,643 |
| Profit after taxes and expenses | $17,453,726 | $22,330,620 |

**Table 11-35**

NPV when softswitch generates 25% more revenue than Class 4

| Class 4 NPV with OCC at 12%, 10%, and 7% | Softswitch NPV with OCC at 12%, 10%, and 7% |
|---|---|
| NPV = ($17,453,726)(1.6901) − $1,085,000 = $28,413,542 | NPV = ($22,330,620)(1.6901) − $225,000 = $37,515,980 |
| NPV = ($17,453,726)(1.7355) − $1,085,000 = $29,205,941 | NPV = ($$22,330,620)(1.7355) − $225,000 = $38,529,791 |
| NPV = ($17,453,726)(1.8080) − $1,085,000 = $30,471,337 | NPV = ($$22,330,620)(1.8080) − $225,000 = $40,148,760 |

income after access fees and Universal Service Fund assessments have been paid. In this model, the retail rate for long distance is $.08 per minute. Access fees and Universal Service Fund payments take $.02 per minute from the Class 4 service provider. The softswitch service provider then enjoys 25 percent greater revenue in this model.

In summary, by not having to pay access fees and Universal Service Fund assessments, the softswitch-enabled service provider enjoys a net present value of over 30 percent more than the Class-4-enabled service provider. This is a compelling reason for long-distance service providers to move to delivering long distance in a phone-to-phone IP service model as an enhanced service provider.

# Large Long-Distance Service Providers

The following pages delve into a "carrier-grade" scenario—comparing the Net Present Values when planning for 36,000 DS0s in one location. Telco strategy over the past few decades has centered on concentrating as many DS0s as possible in one location.

Table 11-36 outlines the gross revenue for large long-distance carriers with 36,000 DS0s. Table 11-37 provides the purchase cost of Class 4 versus softswitch for 36,000 DS0s. Table 11-38 details the costs of purchasing a Class 4 or softswitch. The total benefits and costs of Class 4 and softswitch when purchasing 36,000 DS0s are shown in Table 11-39. Table 11-40 shows the salvage value of Class 4 and softswitch for 36,000 DS0s at 12 percent OCC.

In Table 11-41, the difference in NPV between Class 4 and softswitch is only about four percent in favor of softswitch. This is probably not enough of an advantage, in and of itself, to encourage service providers to select softswitch for their networks. Table 11-42 examines the net present value when the Class 4 solution is $50 per DS0.

**Table 11-36**

Gross revenue for large long-distance carriers with 36,000 DS0s

| DS0s | Minutes/day | Price/minute | Days (2 yrs.) | Total benefit |
|---|---|---|---|---|
| 36,000 | 480 | .03 | 508 | $263,347,200 |

**Table 11-37**

Purchase cost of Class 4 versus softswitch for 36,000 DS0s

| Components of cost | DMS-250 | Softswitch Convergent Networks |
|---|---|---|
| System | $6,220,000 | $1,332,000 |
| Installation | $30,000 | $12,000 |
| Total (P = Co) | $6,250,000 | $1,344,000 |

**Table 11-38**

Costs when purchasing Class 4 and softswitch

| Component of cost | Class 4—DMS-250 | Softswitch |
|---|---|---|
| Collocation rent (annual) | $168,000 (14 racks) | $12,000 (1 rack) |
| Labor | $170,000 (2 engineers) | $140,000 (2 technicians) |
| Maintenance contract | $622,000 | $133,200 |
| IP circuits | $11,000,000 | $11,000,000 |
| Total | $12,395,400 | $11,473,360 |
| For life (j) of 2 years | $24,354,600 | $22,852,640 |

**Table 11-39**

Total benefits and costs of Class 4 and softswitch when purchasing 36,000 DS0s

| Income tax component | Class 4 | Softswitch |
|---|---|---|
| Gross income | $263,347,200 | $263,347,200 |
| Business expenses | $24,354,600 | $22,852,640 |
| Taxable income before depreciation and/or capital gains | $238,992,600 | $ 240,494,560 |
| Salvage value | $782,040 | $631,540 |
| Resale value | $600,000 | $332,000 |
| Book value | $1,022,120 | $295,970 |
| Difference salvage value versus book value | $2,021,459 | $36,030 |
| Capital gains tax at 20% | $404,292 | $7,206 |
| Taxable income | $251,577,080 | $252,098,030 |
| Income tax + capital gains | $88,051,978 + $404,292 = | $85,713,024.06 |
| Profit after taxes and expenses | $163,120,810 | $166,385,005.54 |

**Table 11-40**

Salvage value of Class 4 and softswitch for 36,000 DS0s at 12% OCC

| Salvage value Class 4 (P-S)(A/P, I, N) + Si | Salvage value softswitch |
|---|---|
| ($6,220,000 − $1,350,000)(.5917) + $1,350,000(.12) = $3,043,579 | ($1,332,000 − $332,000)(.5917) + $332,000(.12) = $631,540 |

**Table 11-41**

Net present value
for Class 4 and
softswitch when
purchasing 36,000
DSOs

| Class 4 | Softswitch |
|---------|------------|
| $269,470,481 where N = 2, OCC = 12 | $279,875,297 where N = 2, OCC = 12 |
| $276,876,165 where N = 2, OCC = 10 | $287,429,176 where N = 2, OCC = 10 |
| $288,702,424 where N = 2, OCC = 7 | $299,492,084 where N = 2, OCC = 7 |

**Table 11-42**

Net present value
when the Class 4
solution is
$50 per DS0

| NPV class 4 | NPV softswitch |
|-------------|----------------|
| $274,573,754 where N = 2, OCC = 12 | $279,875,297 where N = 2, OCC = 12 |
| $281,997,793 where N = 2, OCC = 10 | $287,429,176 where N = 2, OCC = 10 |
| $293,853,363 where N = 2, OCC = 7 | $299,492,084 where N = 2, OCC = 7 |

# Summary of Net Present Value Analyses

Table 11-43 gives a summary of the Net Present Values of the preceding scenarios comparing a wide variety of options for the Telco decision maker. It compares the various scales of purchase (480/4,032/36,000 DS0s), the means of acquisition (buy new, buy used, or leased), and the Opportunity Costs of Capital (12, 10, and 7 percent).

The advantages of softswitch in the converging markets outweigh any performance deficiencies. The enabling qualities of softswitch (cheaper, smaller, distributed architecture, and open standards) are strong advantages that outweigh performance deficiencies in the converging market and

**Table 11-43**

Summary of net
present value
analyses

| NPV when purchasing 480 DS0s Class 4 | Softswitch |
|--------------------------------------|------------|
| NPV where N=2, OCC=12=$2,631,675 | NPV=where N=2, OCC=12=$3,172,569 |
| NPV where N=2, OCC=10=$2,717,719 | NPV=where N=2, OCC=10=$3,263,164 |
| NPV where N=2, OCC=7=$2,857903 | NPV=where N=2, OCC=7=$3,407,836 |

*(continued)*

**Table 11-43 (cont.)**

Summary of net present value analyses

| NPV when purchasing used Class 4 480 DS0s | Softswitch |
| --- | --- |
| NPV where N=2, OCC=12=$3,181,020 | NPV=where N=2, OCC=12=$3,114,745 |
| NPV where N=2, OCC=10=$3,269,155 | NPV=where N=2, OCC=10=$3,203,787 |
| NPV where N=2, OCC=7=$3,409,901 | NPV=where N=2, OCC=7=$3,345,997 |
| **NPV when purchasing 4,032 DS0s** | **Softswitch** |
| NPV where N=2, OCC=12=$28,413,542 | NPV=where N=2, OCC=12=$29,415,468 |
| NPV where N=2, OCC=10=$29,205,941 | NPV=where N=2, OCC=10=$30,211,680 |
| NPV where N=2, OCC=7=$30,471,337 | NPV=where N=2, OCC=7=$31,483,163 |
| **NPV when purchasing 4,032 DS0s with Class 4 at $50 per DS0 PLUS VoIP gateway** | **Softswitch** |
| NPV where N=2, OCC=12=$29,248,368 | NPV=where N=2, OCC=12=$29,415,468 |
| NPV where N=2, OCC=10=$30,044,352 | NPV=where N=2, OCC=10=$30,211,680 |
| NPV where N=2, OCC=7=$31,315,470 | NPV=where N=2, OCC=7=$31,483,163 |
| **NPV when purchasing 36,000 DS0s** | **Softswitch** |
| NPV where N=2, OCC=12=$269,470,481 | NPV=where N=2, OCC=12=$279,875,297 |
| NPV where N=2, OCC=10=$276,876,165 | NPV=where N=2, OCC=10=$287,429,176 |
| NPV where N=2, OCC=7=$288,702,424 | NPV=where N=2, OCC=7=$299,492,08 |
| **NPV when purchasing 36,000 DS0s at $50 per DS0 including VoIP Gateway** | **Softswitch** |
| NPV where N=2, OCC=12=$274,573,754 | NPV=where N=2, OCC=12=$279,875,297 |
| NPV where N=2, OCC=10=$281,997,793 | NPV=where N=2, OCC=10=$287,429,176 |
| NPV where N=2, OCC=7=$293,853,363 | NPV=where N=2, OCC=7=$299,492,084 |

make it disruptive to the Class 4 switch. The lower purchase and operating cost of the softswitch compared to Class 4 lowers the barriers to enter the converging market. A precedence has been set with the international long-distance bypass industry. The smaller size and distributed architecture of the softswitch gives the service provider a flexibility not enjoyed with Class 4 switch architecture. The smaller size and significant resulting savings in purchase price and operating costs will drive many service providers to adopt this technology.

Softswitch-equipped service providers enjoy the agility of installing a media gateway in a given market and managing it remotely at great savings over deploying a Class 4 in the same market. Net present analyses indicate that softswitch is only a slightly better value for service providers than Class 4. The exception is where a service provider can generate 25 percent more gross revenue than Class 4 by virtue of offering additional high margin services not possible with a Class 4 switch. Softswitch will prevail in the converging market if for no other advantage than lower costs of acquisition and operations. Also, the ability to roll out new services quickly and enjoy strong margins on those services may prove to be the only reason some service providers deploy softswitch. These advantages may change the market to the extent that any of the performance concerns become insignificant.

Incumbent service providers debating a migration to softswitch would probably not be greatly swayed by single-digit advantages of softswitch over Class 4 when comparing known technology with a new technology. New market entrants may have other considerations. For a new market entrant, a lower cost of entry and exit offered by the softswitch over the Class 4 may be more of a motivating factor than the net present value. In addition, the lower operating cost may also be attractive to the new market entrant. This could lead to more service providers entering the market and diluting the market share of incumbent service providers. In order to really disrupt Class 4 in the converging market, softswitch vendors would be well advised to either cut the sales price and operating costs of their platforms further or emphasize revenue streams through features and applications that Class 4 cannot deliver.

# $0 per port: Subscriber Pays for Access Device Negating "Cost per Port"

Most telecom infrastructure bids are quoted as "dollars per port," a holdover from the days of Class 4 and 5 switches where the competition was for switch vendors to offer the lowest price per DS0, which then assisted the telco in determining their profit per subscriber. Given the rise of broadband and IP handsets and other low-cost media gateways that subscribers are now buying, it is possible that a telco of the not-so-distant future may enjoy a "$0 per port" assessment in offering voice service. A concept that may emerge in softswitch economics will be "cost per busy hour call attempt," as the chief cost components will be softswitch software and the server, not the ports on a switch. When this becomes a reality, TDM service providers will be at a distinct cost disadvantage to VoIP-based service providers. In scenarios where an ISP provides bandwidth and the subscriber obtains voice services from a third party (such as Net2Phone or DialPad VoIP long-distance service providers), VoIP service providers and ISPs are at a distinct advantage in not having to invest in Class 4 switches.

# Economics of Enterprise Softswitch Applications

Corporate interoffice long distance is one of the greatest telecommunications expenses of corporate America. What many chief financial officers, information technology managers, and telecommunications managers probably don't realize is that their firm is paying for *two* networks: one data and one voice. Furthermore, they probably don't realize that they could be routing their voice traffic over their data network and saving the firm and its shareholders millions annually.

Any firm can greatly reduce if not eliminate their corporate interoffice long-distance expenses by installing IP *private branch exchanges* (PBXs) on their corporate networks (see Tables 11-44 and 11-45). A business can route its interoffice voice traffic (local and long distance) over its existing corporate *wide area network* (WAN). This potentially eliminates the businesses' interoffice long-distance bills.

# Non-Interoffice Free Long Distance

When much of a firm's long-distance traffic is to the same local calling areas (extensions 303, 408, 212, 612, and so on), using the off net dialing function of a VoIP gateway switch installed in those local calling areas will enable employees to get a dial tone in those local calling areas or overseas and incur no long-distance charges (see Table 11-46). Table 11-47 provides a worksheet comparing long-distance costs to leasing VoIP gateways in various *local access and transport areas* (LATAs).

**Table 11-44**

*Worksheet for determining return on investment for IP PBX*

| | | | |
|---|---|---|---|
| Office A monthly interoffice long-distance bill | $_____ | VoIP gear for office A | $_____ |
| Office B monthly interoffice long-distance bill | $_____ | VoIP gear for office B | $_____ |
| Office C monthly interoffice long-distance bill | $_____ | VoIP gear for office C | $_____ |
| Office D monthly interoffice long-distance bill | $_____ | VoIP gear for office D | $_____ |
| Subtotals | $_____ | | $_____ |
| Federal, state, and local taxes | $_____ | | $_____ |
| Totals | $_____ | | $_____ |

Return on Investment = the number of months for interoffice long-distance costs to equal cost of IP PBX gear, that is, the number of times column A goes into column B.

**Table 11-45**

*Worksheet comparing interoffice long-distance and leasing an IP PBX*

| | | | | | |
|---|---|---|---|---|---|
| Office A monthly interoffice long-distance bill | $_____ $_____ | Monthly lease payment for VoIP gear for Office A | $_____ | Monthly savings at Office A | $_____ |
| Office B monthly interoffice long-distance bill | $_____ | Monthly lease payment for VoIP gear for Office B | $_____ | Monthly savings at Office B | $_____ |
| Office C monthly interoffice long-distance bill | $_____ | Monthly lease payment for VoIP gear for Office C | $_____ | Monthly savings at Office C | $_____ |
| Office D monthly interoffice long-distance bill | $_____ | Monthly lease payment for VoIP gear for Office D$ | $_____ | Monthly savings at Office D | $_____ |
| Subtotals | $_____ | | $_____ | | $_____ |
| Taxes | $_____ | | $_____ | | $_____ |
| Totals | $_____ | | $_____ | | $_____ |

**Table 11-46**

Worksheet for comparing long distance to acquiring VoIP gateways in frequently called LATAs

| | | | |
|---|---|---|---|
| Monthly cost to call local calling area A | $_____ | VoIP gear for local calling area A | $_____ |
| Monthly cost to call local calling area B | $_____ | VoIP gear for local calling area B | $_____ |
| Monthly cost to call local calling area C | $_____ | VoIP gear for local calling area C | $_____ |
| Monthly cost to call local calling area D | $_____ | VoIP gear for local calling area D | $_____ |
| Subtotals | $_____ | | $_____ |
| Federal, state, local taxes | $_____ | | $_____ |
| Totals | $_____ | | $_____ |

**Table 11-47**

Worksheet comparing long-distance costs to leasing VoIP gateways in various LATAs

| | | | | | |
|---|---|---|---|---|---|
| Monthly cost to call local calling area A | $_____ | Monthly lease for VoIP gear in local calling area A | $_____ | Monthly savings to call local calling area A | $_____ |
| Monthly cost to call local calling area B | $_____ | Monthly lease for VoIP gear in local calling area B | $_____ | Monthly savings to call local calling area B | $_____ |
| Monthly cost to call local calling area C | $_____ | Monthly lease for VoIP gear in local calling area C | $_____ | Monthly savings to call local calling area C | $_____ |
| Monthly cost to call local calling area D | $_____ | Monthly lease for VoIP gear in local calling area D | $_____ | Monthly savings to call local calling area D | #_____ |
| Subtotals | $_____ | | $_____ | | $_____ |
| Taxes | $_____ | | $_____ | | $_____ |
| Totals | $_____ | | $_____ | | $_____ |

# Implications for Developing Economies

As business *is* communications, a robust telecommunications network is probably the most important element of economic development. Teledensity is the number of telephone lines per hundred persons. The United States enjoys a teledensity of 66 percent. Mexico, in comparison, has a teledensity of 13.7 percent with government initiatives to almost double that by 2006.[16] Softswitch technologies will be critical in achieving that goal. In the past, teledensity had been constrained by the extreme expense of Class 4 and 5 switch technologies both in acquisition and OAM&P. Softswitch technologies, given their lower costs, simplicity of operation, and ease of integrating custom services, will accelerate teledensity in developing economies around the world.

An excellent case in point is Protel, a Mexican ISP and long-distance company. Protel has built an IP backbone that links the major population centers of Mexico. On that IP backbone, Protel routes VoIP long distance at about five million minutes of voice traffic monthly. Protel will expand this service to 35 cities by 2003.

When Protel decided to build out its VoIP network, it originally focused on reducing the cost of long-distance services for its commercial subscribers. Today Protel is expanding its IP network to offer customers enhanced services such as dedicated Internet access, wholesale dial, *virtual private networks* (VPNs), *Multiprotocol Label Switching* (MPLS) VPNs, and long-distance VoIP.

By rolling out this extensive long-distance network, Protel has introduced competition into the Mexican long-distance market. This inevitably has driven down the cost of long-distance services. Lower long-distance costs spur greater trade within the 35 cities served by Protel. Greater trade is a boost to economic development in Mexico.

The Protel network includes Cisco AS5300 voice gateways and Cisco SC2200 signaling controllers (in essence, a softswitch) for SS7 interconnections with the Mexican PSTN throughout Mexico.[17] Although these platforms

---

[16]Comisión Federal de Telecomunicaciones. "Communications Sector Program 2001–2006." A planning document from the Mexican government, 2001. Available at www.cft.gob.mx/html/presidencia/communicationsectorprogram.pdf.

[17]Cisco Systems. "i-Next, first IP multi-service network in Latin America, expands its reach with Cisco VoIP networking equipment." A press release available at http://newsroom.cisco.com/dlls/corp_082701.html.

may not be considered carrier grade by North American carriers, they are excellent examples of Christensen's definition of disruptive technology in that they are cheaper, simpler, smaller, and more convenient to use than the Class 4 switches they replace. The chief reason teledensities of emerging markets lag behind that of the developed world is that the economies of Class 4 and Class 5 switches are too expensive for those developing nations. It is difficult to build an economic model for these countries that would support switches that cost tens of millions of dollars each. If softswitch technologies costing 30 percent of the cost of a Class 4 or 5 switch were to be substituted for the Class 4 or 5 switch, it is much easier to develop economic models that will support an expanded telecommunications infrastructure (especially converged voice and data). Vendors of softswitch-related technologies who have found their North American markets to be depressed at the time of this writing would be well advised to seek out emerging markets.

# Economic Benefits of Converged Networks

In their April 2001 white paper, "The $500 Billion Opportunity: The Potential Economic Benefit of Widespread Diffusion of Broadband Internet Access," Robert Crandall and Charles Jackson point to an economic benefit of $500 billion per year for the American economy if broadband Internet access were to be as ubiquitous as landline phones.

Any new product or product improvement creates benefits for both consumers and producers. Consumers gain because they are able to purchase a new or improved product that was previously unavailable. They consume it up to the point at which the marginal value of the product is equal to its price. In the case of typical broadband services, consumers either subscribe to the service or they do not. As the uses of broadband multiply, the value to subscribers rises far above the monthly subscription price. This is the consumer surplus from the innovation.

Producers of new services that rely on broadband, of products used in conjunction with broadband service, and even of the broadband service itself also gain from the greater diffusion of broadband. They bid resources away from other sectors of the economy and earn returns over and above those available elsewhere until the marginal value of each type of resource is equalized across all alternatives. The *producer surplus* that is generated by inframarginal sales is a real benefit to producers and therefore to the economy.

In this section, we attempt to provide a rough estimate of the likely long-term gains to the economy: the sum of consumer and producer surplus generated by the widespread diffusion of broadband access. By long term, we mean a time period sufficient for broadband to become virtually ubiquitous, given the appropriate policy environment, a time period that could stretch out to 25 years or even more.

## Broadband Access and Telephone Services

What is the big deal behind softswitch if it doesn't have some enduring impact on society? As a key network element in the converging/converged network, softswitch will improve the lives of the subscribers it serves. Ubiquitous broadband in the United States could bring a $500 billion annual benefit to the American economy. Similar benefits could be realized in other nations. Incorporating voice and voice features on that broadband network is absolutely essential in building a converged network.

Soon the demand for broadband will reflect not only the growing potential uses of the Internet, but also the prospect for using these broadband connections to obtain voice telephone services currently provided over a narrowband connection. The use of broadband access to carry voice (ordinary telephone calls) as well as data will deliver substantial savings to consumers that are not captured in current demand estimates.

A variety of options for VoIP are available in the residence or small business. An IP handset connects directly to the IP network via an Ethernet connection. Media gateways for the residence or small business range from one to eight ports that connect analog handsets to the IP network via a standard RJ-11 telephone jack. The PC, loaded with the appropriate software, can also be used with the speakers and microphone resident on the sound card in the PC. For those who find using a PC to be an anthropological challenge, an analog handset can be attached to the PC with the feel of a conventional telephone.

A broadband connection can support several voice connections; the exact number depends on the speed of the connection and the degree of compression of the voice signal. The current structure with two networks in the home (a voice and an IP network) and two connections to the outside world (a narrowband analog connection and a high-speed digital connection) is inefficient. However, the transition to Internet telephony will take a few years. The existing inside wiring, designed for voice systems, will probably be replaced by *wireless fidelity* (Wi-Fi) applications. The analog telephones must be replaced or connected to digital adapters, and difficult issues of addressing, numbering, and directory services will arise. Broadband connections must become highly reliable, and the quality of packet telephone calls must improve.

The cost savings from integrated access will be significant. Reliable Internet telephony would eliminate the need for second or third lines in households for teenagers or fax machines. The FCC has estimated that the average household spent $55 per month on local and long-distance telephone service in 1999, and 0.289 additional lines existed for each household with telephone service.

Within a few years, broadband access will permit consumers to substitute other services for the ones that now cost $55 per month. The FCC estimates that the average residence spends $34 per month for local telephone service and $21 for long-distance telephone service. Part of that local telephone service cost is for the loop that is used for the broadband service. Consumers continue to incur most of those loop costs when broadband service is used, but they avoid the cost of the analog line card, the voice switch, and the voice transmission lines.

IP telephony should lower the costs of both local and long-distance telephone service while providing residences with the equivalent of several telephone lines. We estimate that such savings could average $25 per month per household. In addition, households with broadband service would get the equivalent of multiple voice telephone lines. We estimate that this additional service or option of service could be worth $10 per month to the average household.[18] Thus, in the longer run (say, a decade from now), broadband access could deliver voice communication benefits of about $35 per month, or $420 per year, to the average household with telephone service. If we assume that 122.2 million households have telephone service, these benefits would total $51.4 billion per year, assuming no growth in voice usage. The actual value could be much higher.

The substantial economic benefits (principally savings from expenditures on telephone service) created by providing multiple services over a high-speed line almost cover the cost of a high-speed line; we have estimated that benefits of $35 per month are created by a broadband connection that costs

---

[18]The FCC's numbers indicate that the average household with telephone service has 1.289 access lines and pays local service fees of $34 per month. Assuming that all lines cost the same (which is not quite right but is reasonable), the average household with telephone service in 1999 paid $7.62 per month for additional line service. If those households without a second line today place an average value of no more than $3.40 per month for a second line of service, then the average household will value a second line at $10 per month or more.

$40 per month. These savings are one reason why it is reasonable to expect that the number of households with phone service today will ultimately be the same number as those with high-speed access services.[19]

# Conclusion

To reemphasize Christensen's definition, a disruptive technology is cheaper, simpler, smaller, and more convenient to use. This chapter demonstrates *how* softswitch solutions match this definition. They are cheaper in purchase price, OAM&P, power draw, real estate, and so on. They are simpler in that they do not require a small army of engineers to operate them (especially SIP-based IP PBXs). They are smaller in that they take as little as one-thirteenth of the real estate required of a circuit switch of similar scale. Also, they can scale down to single-port devices. Finally, as the examples of Protel and IP PBXs demonstrate, they are becoming increasingly more convenient to use.

---

[19]Another way to look at it is to consider the engineering design. Why have two line cards at the telephone central office—one in the voice switch and one in the DSLAM—if the line card in the DSLAM has roughly 10 times the capacity of the line card in the voice switch? If we were building the telephone network from scratch today, broadband access connections would be the natural building block.

# Is Softswitch Deconstructive, Disruptive, or Both?

Since the introduction of *Voice over IP* (VoIP) in the last decade, most of the debate comparing VoIP to the *Public Switched Telephone Network* (PSTN) or softswitch to the Class 5 switch focuses on a DS0-to-DS0 replacement of the incumbent technology by the new technology. It has also consisted of a feature-by-feature comparison followed by a component-by-component comparison. This is an inaccurate means of comparing the two technologies. Much of the terminology focuses on softswitches as Class 4 or 5 replacements. A better perspective is to view the new technology as superceding as opposed to replacing Class 4 or Class 5 switches. This chapter will focus on two concepts currently in vogue regarding new technologies superceding incumbent technologies: deconstruction and disruption.

In their 2000 book entitled *Blown to Bits*[1], Phillip Evans and Thomas Wurster explore how certain industries have been "deconstructed" by the Internet. That is, the emergence of information or services available via the Internet has caused firms to lose sales and market share if not their entire business due to the emergence of new technologies. Examples of those industries include travel agencies, retail banks, and automobile retailers. The following pages will investigate the potential deconstruction of the North American telecom industry by Internet-related telephony applications.

Clayton M. Christensen in *Innovator's Dilemma* (referred to earlier) describes how some well-managed firms can find themselves disrupted by new technology that is cheaper, simpler, smaller, and more convenient to use, known as disruptive technology. Has the PSTN or major components of the PSTN been disrupted by softswitch technologies? This chapter will study both of these concepts relative to the impact of softswitch on the current telecommunications markets.

# Deconstruction

The telecom sector in recent years has been deconstructed by technologies that are Internet related, if not by the Internet itself. Long-distance bypass using VoIP, as described earlier in this book, is a good example of such a technology. The delivery of telephony features to a voice service via the *Internet Protocol* (IP) would also be an example of deconstruction of the telecom service provider industry by an Internet-related technology.

---

[1]Evans, Philip, and Thomas Wurster. *Blown to Bits*. Boston, MA: Harvard Business School Press, 2000.

## Deconstruction of Service Providers

Incumbent telephone service providers are being deconstructed as their market share shifts from their networks to IP-based networks and applications. As incumbent service providers lose revenue, they have less money to spend on infrastructure. This inevitably affects the vendors that supply incumbent service providers with legacy platforms (Class 4 or 5 switches and *private branch exchanges* [PBXs]). Christensen's *Innovator's Dilemma* describes "value networks"[2] as consisting of, in this case, service providers and the vendors that service them. As the service providers see their market shares and resultant revenues fall off, their vendors will also be adversely affected.

Potential deconstruction of the telecom industry by Internet-related technologies focuses on service providers and the vendors that provide their infrastructure. A discussion of carriers is in order. Perhaps the most vulnerable are long-distance providers such as the Big Three (WorldCOM, AT&T, and Sprint). Seventy percent of corporate telephone traffic is employee to employee, that is, office to office. As this traffic moves to the corporate *wide area network* (WAN), the long-distance traffic, for example, is almost free (the expense is the bandwidth and new VoIP infrastructure at the very edge of the network). To dramatize this point, if 70 percent of corporate long distance migrates away from the Big Three and onto the WAN, the Big Three are severely deconstructed by Internet-related technology (VoIP). The new IP PBXs, especially those that are *Session Initiation Protocol* (SIP) based, to quote Christensen, are "cheaper, simpler, smaller, and more convenient to use" than legacy PBXs. The inclusion of SIP on Microsoft's Window's XP, as another example of potential deconstruction, enables every Windows-XP-equipped PC to be a telephone. This could shift 245 million XP users to be long-distance subscribers to IP-based service providers and/or the corporate WAN in the next 2 to 3 years. No phone company in the world has 245 million subscribers.

In order to understand how softswitch, particularly softswitch as a Class 4 replacement, could deconstruct the long-distance market, it is important

---

[2]A value network, as defined by Christensen, incorporates both the physical attributes of a product or system as well as the associated cost structure (as typically measured by the gross margin). This cost structure includes all business-related costs (research, engineering, development, sales, and marketing) necessary to sustain a business. A value network also includes all the critical upstream and downstream value chain participants. Tightly intertwined value networks prevent disruption by giving incumbents an easy avenue to co-opt potentially disruptive technologies by cramming them into their existing business model.

to study what has already transpired in the North American long-distance market as barriers to entry have been lowered. The introduction of softswitch will serve to lower those barriers further. As described earlier in this book, the lower purchase and *Operations, Administration, Maintenance, and Provisioning* (OAM&P) costs of softswitch relative to Class 4 will lower the barriers to both entry and exit for the North American long-distance market. This will encourage more new market entrants and accelerate the loss of market share of incumbent long-distance service providers.

Those most likely to enter the long-distance market are telecommunications service providers that already have customers and some of the necessary infrastructure. These potential market entrants include cable TV service providers, *Internet service providers* (ISPs), wireless service providers, application service providers, and *Regional Bell Operating Companies* (RBOCs). It should be noted that the introduction of softswitch in the long-distance market potentially accelerates a trend that has been in effect over the last few years. The Table 12-1 details the decline in market share of the top 20 international long-distance service providers and the growth in market share enjoyed by new market entrants.

These new market entrants will take market share away from the incumbent long-distance service providers that use the more expensive and less flexible Class 4 switch. Eventually, competition will drive the price per minute for long distance downward. Incumbent long-distance service providers that use the Class 4, which is more expensive to operate than softswitch, will find themselves less competitive than their softswitch-equipped competitors. This could potentially deconstruct the incumbents and drive them out of business.

One long-distance service provider that is potentially disrupted by softswitch-enabled new market entrants is AT&T. Over the past few years,

**Table 12-1**

Market share comparisons in international long distance, 1990–2000

| Indicator | 1990 | 1995 | 2000 |
|---|---|---|---|
| Revenues in international traffic ($ billions U.S.) | $37 | $55 | $70 |
| Top 20 carriers share of world traffic | 86% | 72% | 50% |
| Market share of new carriers | 1% | 5% | 31% |

Source: TeleGeography

---

[3] "AT&T To Take Biggest Hits in Long-Distance Fight." *Lightwave*, Vol. 14 Issue 10, September 1997. pg. 118. Available online at http://lw.pennnet.com/Articles/Article_Display.cfm?Section=Archives&Subsection=Display&ARTICLE_ID=33824

AT&T has lost market share to new market entrants.[3] That trend could now be accelerated, given an ease of market entry and exit made possible by softswitch technologies. According to a report issued by the *Federal Communications Commission* (FCC), AT&T's share of revenues in the long-distance market decreased to nearly 40 percent by 1999, just after the company was split into several independent baby Bells in 1984. Other long-distance carriers, in particular WorldCOM and Sprint, took advantage of this fact, trying to gather more than 25 percent and 10 percent, respectively, of the share of toll revenues by the end of 1999. On the other hand, more than 700 smaller long-distance providers had the remaining 25 percent market share.[4] Those 700 smaller long-distance carriers potentially can take more market share from AT&T if they adopt softswitch technologies that can cut their overhead and enable them to offer services not available from AT&T, MCI, or Sprint.

Evans and Wurster's book debates the contradiction of reach versus richness. A business plan based on reach offers a product that is highly commoditized and easily distributed. A product based on richness is much more rich in information and customer contact. In the case of telecommunications, long-distance service is an example of a business plan based on reach as long-distance minutes have become a commodity (which might explain why over 700 long-distance providers exist in the United States). Richness implies a deeper customer-service provider relationship with a greater range of services tailored to the individual consumer. Such a relationship might exist between a large enterprise customer and a service provider, but not with a residential or small business service provider.

A key element of deconstruction states that the chief competitor to any business is their customer. This is focused on the question, "What does the business provide the customer that the customer cannot do for themselves?" An enterprise moving their voice traffic to their WAN would be an example of customers hauling their voice traffic for themselves, negating the need for a voice service provider. As more enterprises move their voice traffic (commoditized minutes or T1 lines) to their WAN, service providers will have to shift their focus to the richness of the relationship with their customer. One approach by service providers for developing a richer relationship with their customers could be to offer a wider range of custom features not available from competing service providers or not easily done by the customer. This will only be possible by softswitch-equipped service providers.

---

[4]"Long-distance market share of AT&T decreased to 40% in 1999." Mobile News, January 26, 2001, http://uk.gsmbox.com/news/mobile_news/all/28619.gsmbox.

# Vendors

A quick review of telecommunication vendors shows a serious downturn in their sales and stock price over the last two years. Traditional telecommunications switch vendors Lucent technologies and Nortel are good examples. Formerly, these two titans of the industry dominated the North American switching market in Class 4, Class 5, and PBX switching platforms. Internet-related technologies are potentially deconstructing the current North American switching industry.

Although these vendors will not disclose revenue figures for specific platforms, a brief review of announced sales reveals no significant sale of Class 4 or 5 switches in the past two years. Sales of these switches are off because many service providers have expressed a reluctance to invest in Class 4 or 5 switches as softswitches enter the market. Softswitch or Internet-related technologies may be deconstructing the telecom vendor industry via a form of psychology. Large service providers have made it no secret that they are not going to invest in any Class 4 or 5 switches as they wait for softswitches and Internet-related technologies to emerge. Given the traditional long depreciation schedules of traditional switches, decision makers at the incumbent telcos must plan in terms of future decades. In short, they must have a vision for the telecom infrastructure against which they will be competing 10 years in the future. As described in earlier chapters, softswitch offers a flexibility in delivering features not possible on Class 4 and 5 switches. Service providers do not want to find themselves at a distinct disadvantage to their competitors in the future once softswitch technology matures and becomes more widely installed. Among RBOCs and *interexchange carriers* (IXCs) is the decision to wait and see.

Suffice it to say that if the large carriers are strapped for cash, they are not spending money on new infrastructure. In addition, a wave of mergers and acquisitions in the industry has also made it wise to avoid *cap ex*, or capital expenditure, which makes the balance sheet of the carrier look better to an acquiring party. Some carriers at the time of this writing are so deeply in debt that they have been labeled as being credit challenged, which negates vendor financing of the sales of any new products.

A snapshot of new technologies in the telecom industry points to other deconstructive forces. VoIP can be delivered in increments of one IP phone, which negates the need for a big centralized switch. Failing that, the deployment of four-port or eight-port media gateways that connect to the IP network for VoIP also negates the need for a large centralized switch. Vendors of these smaller platforms are numerous with new market entrants occurring frequently due to a lower barrier to entry in the vendor market. Many of these products originate in Asia where manufacturing costs are relatively low.

Mainstream Class 4 vendors are Lucent and Nortel. Lucent's fiscal year for 2001 ended September 30 with an annual loss of $16.2 billion. Nortel networks reported an annual loss to date on September 30, 2001 of $4 billion with another quarter to go to finish their fiscal year. Although these are the two mainstream providers of Class 4 switches in North America and they have been in business a long time, these financial losses point to serious problems for those firms. The chief products of both firms were circuit-switched telephone switches in the Class 4 and 5 markets as well as PBX. The challenge for these mainstream vendors is to produce a product line that can compete favorably in the converging market and be well positioned for the converged market.

# Disruption of the Legacy Telecommunications Value Network

"Telecom meltdown" strikes far beyond the woes of the carriers. It extends to the vendors of the carriers and all parties that derive their revenues from the carriers. When carriers cease to make money, they cease to spend money on new platforms and services. Vendors used to multibillion dollar sales figures then find themselves losing hundreds of millions of dollars per quarter. Many find themselves selling the wrong products in the wrong markets.

## Disruption in Class 4 Market

Table 12-2 lists the pros and cons of softswitch being a competitor to Class 4 switches. Although many argue that softswitch will disrupt Class 4 in the converging market, it is also argued that softswitch will not displace softswitch in the converging market. The following pages will weigh those arguments.

**Class 4 Disrupted by Softswitch**  The following section will present arguments in defense of softswitch over Class 4. Those arguments highlight the technical, financial, and regulatory advantages of softswitch. The most

**Table 12-2**

Pros and cons:
Is softswitch
disruptive?

| Pros | Cons |
| --- | --- |
| Given the technical performance parameters of softswitch matching those of Class 4, financial advantages such as the purchase and operation cost of softswitch are driving the deployment of softswitch and the replacement of Class 4 by incumbent service providers to save on OAM&P costs. | Mainstream vendors, despite annual losses in the billions, survive and sell softswitch to mainstream service providers with no part of value network disrupted. |
| Bandwidth glut makes VoIP more attractive. New market entrants are encouraged by the economics of an all-IP network and invest in softswitch. | Softswitch is only a client-server version of the mainframe Class 4 and thus constitutes a sustaining technology as opposed to a disruptive technology. |
| Given that softswitch matches the technical performance parameters of Class 4, softswitch proceeds to exceed Class 4 performance parameters, making it the obvious choice for upgrades by service providers, both incumbents and new market entrants. | Mainstream (Class 4) vendors slash the purchase price of Class 4 switches to either match or beat softswitch. |
| A softswitch costs less than a Class 4 switch in terms of purchase price and OAM&P. This constitutes a lower barrier to entry into the long-distance market. This has the effect of encouraging new market entrants (ISPs, cable TV companies, power companies, and so on) to expand their services to include long-distance services. In additon, many RBOCs can now offer long-distance services in their operating regions. Softswitch will allow RBOCs a cost-effective means of rolling out long-distance service in those areas where existing infrastructure will not enable long distance. | Mainstream carriers are rushing to the converged market with the deployment of Class 5 softswitch solutions, accelerating movement toward the converged market and negating the comparison of Class 4 to softswitch in the converging market. |
| Softswitch has the potential to offer features and applications not available via Class 4 technology. This means more high-margin services that service providers can offer their subscribers. | The net present value analysis does not strongly support the advantages of softswitch over Class 4. Service providers will not invest in this unproven technology when it shows only a 3 percent advantage in the net present value over Class 4. |
| Regulatory issues related to access fees and the Universal Service Fund are driving long-distance traffic to phone-to-phone, IP-enhanced service provider models that are excluded from access fees and Universal Service Fund costs. Service providers will deploy softswitch solutions to take advantage of avoiding these costs. | The FCC may reverse itself and impose access fees and USF assessments on IP phone-to-phone calls, eliminating the advantages of enhanced service providers. |

**Table 12-2 cont.**

Pros and cons:
Is softswitch
disruptive?

| Pros | Cons |
| --- | --- |
| Mainstream Class 4 vendors have not succeeded in achieving a significant market share with their Class 4 replacement softswitch solutions, leaving the market to start up softswitch vendors. | Mainstream long-distance service providers are slashing their rates to the point of losing money. This is done to deny profitable margins to new market entrants that use softswitch and thus inhibit the growth in new market entrants and hinder the sales of softswitch. |
| The operating costs of Class 4 cannot be reduced. Class 4 will always require 10 or more times as much rack space as softswitch and draw more power than softswitch, making Class 4 always more expensive to operate than softswitch. | Mainstream long-distance service providers with Class 4 switches, armed with excess capacity on their existing networks, are slashing their rates to deny new market entrants any opportunities at profitability. |
| The growth in enterprise VoIP telephony demands small, enterprise-grade softswitch solutions such as IP Centrex and IP PBX, leading to the dilution of the business long-distance market and the erosion of mainstream long-distance service providers' market share. | Internet depression denies capital to new market entrants, as discussed in Michael J. Mandel's *The Coming Internet Depression*. Incumbents with legacy Class 4 switches continue to hold market share. Incumbents do not invest in new plants due to capital crunch. With new market entrants not emerging due to capital crunch and capital denied to softswitch vendors, softswitch vendors go out of business. |
| The simplification of VoIP made possible by new media gateways and softswitch applications allows alternative service providers (*Community Antenna Television* [CATV], ISPs, RBOCs, municipal *metro area network* [MAN] operators, and applications service providers) to offer long-distance service and disrupt mainstream service providers. | |
| The events of September 11, 2001 (the destruction of switching facilities, leaving 300,000 subscribers without service) drive the fear of the loss of central office services and promotes a move to softswitched IP networks, driving a growth in softswitch sales. | |
| Under Section 271 of the Telecommunications Act of 1996, RBOCs gain permission to offer long distance in their regions. The need is growing to deploy switching gear fast. Softswitch provides that advantage (Qwest and SONUS are examples of this). | |
| Mainstream service providers fear mainstream Class 4 vendors will discontinue Class 4 product lines, leaving them without technical support for their existing Class 4 switches. They are choosing softswitch to protect themselves. | |

simple argument in favor of softswitch is the financial advantages, as seen in the purchase and operating costs of softswitch. The previous chapter compared the real-world purchase and operating costs of softswitch and Class 4. In almost every instance, softswitch is cheaper to purchase and operate than Class 4. This proves to be attractive to new market entrants and to incumbent service providers seeking to upgrade their existing infrastructure.

The chief argument service providers have had against softswitch in the past is that it does not match the performance parameters of Class 4. Softswitch now matches or exceeds the performance parameters of Class 4 in terms of reliability (five 9s), scalability (100,000-plus DS0s and over 800,000 *busy hour call attempts* [BHCAs]), *quality of service* (QoS, 4.0 plus *mean opinion scores* [MOSs]), and the features and applications (transmits all features and applications that a Class 4 does in addition to new IP-centric features).

Given that softswitch matches or exceeds the performance parameters of Class 4 and that it is less costly to purchase and operate, new market entrants to the converging long-distance market will want to acquire and deploy softswitch in place of Class 4. These lower costs equate to lower barriers to entry and exit for those new market entrants. Service providers such as ISPs, cable TV providers (CATVs), application service providers, nonfacilities-based long-distance resellers, power companies, and broadband service providers are now enabled to offer long-distance service to their new and existing customers.

In addition to those new market entrants, RBOCs are gradually entering the long-distance market in their respective regions. To do so, they must satisfy conditions established in Section 271 of the Telecommunications Act of 1996. The Act requires RBOCs to meet a federally mandated 14-point checklist before they offer long distance in their operating region. The checklist includes interconnection and collocation; unbundled network elements; poles, ducts, conduits, and rights of way; loops; transport; switching; 911, directory assistance, and operator services; white page directory listings; number administration; signaling and databases; number portability; dialing parity; resale; and reciprocal compensation before they can offer long distance within their own operating regions.

As the RBOCs succeed in meeting those requirements and make plans to roll out long-distance service in their operating areas, they will want to use equipment that is competitively priced and can be quickly deployed. Softswitch, given that all other factors are equal (scalability, reliability, QoS, and features), is a more logical choice in this role than Class 4. If multiple RBOCs were awarded the opportunity to offer long-distance service throughout the United States in, say, one year, there would be a rush to deploy infrastructure to provide those services. This would be a boon for softswitch and the argument in favor of it over Class 4 switches.

In addition to being less expensive to acquire and operate, softswitch has the potential to generate more high-margin features than Class 4. Softswitch can transmit all the same features that Class 4 does. Open standards that enable expanded interfaces with applications hosted on an application server expand the number of features a softswitch-enabled service provider can offer. This allows that service provider to differentiate themselves from other carriers, especially Class-4-enabled service providers who physically cannot deliver the same services.

Net present analyses in Chapter 5, "SIP: Alternative Softswitch Architecture?" show that where all other things are equal and the softswitch-enabled carrier is generating 25 percent more revenue than they would with a Class 4, the net present value is about 25 percent higher than it would be if the carrier were using a Class 4 that could not generate the 25 percent higher revenue due to the Class 4's incapability to offer the additional revenue-generating features. If more carriers select softswitch over Class 4 for this reason, Class 4 vendors will lose sales and be disrupted.

Long-distance service providers that haul their traffic via IP are exempt from access fees and levies for the Universal Service Fund. Although VoIP can be delivered via an appropriately equipped Class 4, service providers seeking an all-IP solution for maximum efficiency will select softswitch over Class 4. Service providers will deploy softswitch solutions to take advantage of avoiding these costs. If the FCC were to reverse itself, this advantage for IP telephony and softswitch would disappear.

Enterprise VoIP is growing in popularity. Many businesses seek to reduce or eliminate their long-distance costs. One way to do so is to route their interoffice long distance over their corporate WAN. This is traffic that formerly was routed via a mainstream long-distance carrier using Class 4. Assuming this trend continues, Class 4 in the form of a long-distance carrier using Class 4 is disrupted.

In the aftermath of the attack on the World Trade Center, corporate subscribers will demand greater security for their voice services. The World Trade Center and buildings around them contained both Class 4 and Class 5 switches. The destruction of those towers left hundreds of thousands of business subscribers without service for weeks. One possible alternative is to route corporate interoffice traffic over the corporate WAN, so that in the event of the destruction of the local central office, it is possible to have backup telephone service via the corporate WAN.

VoIP presents both a disruptive threat and a potential opportunity for traditional IXCs who provide advanced telecommunications services to businesses. If an IP network can replicate the network services, new *Internet telephony service providers* (ITSPs) could provide IP-based services such as call center management and *voice virtual private networks* (voice VPNs) that would threaten traditional IXC's high-margin service offerings. These offerings would leverage the inherent flexibility of an IP-based solution and would allow true one-stop shopping for businesses.[5]

**Class 4 Not Disrupted by Softswitch**   Just as evidence supports the thesis that softswitch disrupts Class 4, an argument can be made that softswitch is not disruptive to Class 4. The greatest threat to this claim has little to do with the technical differences of softswitch over Class 4.

The greatest threat to this argument is that mainstream vendors of Class 4 who also offer a softswitch product and sell largely to mainstream carriers will discontinue their Class 4 products and sell their softswitch solutions to mainstream service providers. In this scenario, no Class 4 vendor or incumbent long-distance service provider is disrupted. In this case, softswitch becomes a sustaining technology. Another potential argument is that softswitch is the client/server version of a Class 4 switch that is based on mainframe technology. In this case, softswitch could be considered a sustaining technology, providing no member of the value network is disrupted.

Mindful of lost sales to softswitch, mainstream (Class 4) vendors may slash the purchase price of Class 4 switches to either match or beat softswitch. Some vendors are reportedly selling switches for $50 per DS0 in some Asian markets. This potentially negates the low purchase price advantage of softswitch. If Class 4 vendors follow this strategy, then the same conditions that apply to softswitch as regards a lower barrier to entry and exit will apply to Class 4 and that advantage is lost to softswitch.

Net present value analyses do not strongly support long-term financial advantages of softswitch over Class 4. Service providers will not invest in this unproven technology when it shows only a three percent advantage in the net present value over Class 4. Class 4 has 25 years of experience in the long-distance market. Incumbent carriers may take a wait-and-see attitude toward the new technology.

Carriers that offer phone-to-phone IP telephony are exempt from paying access fees or paying into the Universal Service Fund. For this argument, both the softswitch and Class 4 service providers enjoyed this advantage

---

[5]Christensen, Clayton, Scott Anthony, and Erik Roth. "Innovation in the Telecommunications Industry—Separating Hype From Reality." A working paper from the Harvard Business School, October 2001.

over non-IP service providers. This freedom from paying those fees is a significant advantage for carriers offering VoIP services and drives the market demand among carriers to deploy VoIP solutions, including softswitch. Those carriers would be most likely to select an all-IP solution (softswitch) over a partial IP solution (Class 4). If the FCC reversed itself and imposed access fees and Universal Service Fund assessments on IP phone-to-phone calls, this would eliminate the advantages of enhanced service providers and diminish the attractiveness and market share of softswitch.

With lower operating costs, VoIP carriers can potentially take market share and profits from mainstream long-distance service providers. In response, mainstream long-distance service providers may lower their rates to the point of losing money to deny profitable margins to new market entrants that use softswitch and thus inhibit the growth of new market entrants. Mainstream service providers may be enabled to slash their prices due to excess capacity on their existing networks and deny new market entrants any opportunities at profitability. This will further erode the demand for new service providers in the long-distance market and as a result the demand for softswitch.

In addition to cutting their long-distance rates, incumbent service providers may adopt an anticonvergence mentality to preserve their market share. An Internet depression would stifle investments in both softswitch and new long-distance service providers that would have deployed softswitch. Incumbents could adopt a strategy of discontinuing investments in new technologies as they face no real competition from new market entrants armed with softswitch and other next-generation technologies. Examples of this are Sprint's discontinuation of ION and slow roll-out of its Sprint Broadband, a wireless broadband Internet service, and SBC's slowing deployment of Project Pronto, a fiber-to-the-home broadband Internet service. The demise of *digital subscriber line* (DSL) providers Rhythms and JATO, for example, have enhanced the market position of incumbent service providers such as Qwest, SBC, and Verizon. With fewer IP services being implemented, softswitch would have a less fertile market in which to operate.

An Internet depression would have the effect of further curtailing investments in softswitch vendors and their associated value network would no longer include third-party software vendors that develop "killer apps" to go with softswitch in order to generate new high-margin services. With existing service providers avoiding cap ex on infrastructure (namely switching equipment) and thus not investing in softswitch and other next-generation equipment, softswitch vendors would be denied sales revenue, driving some softswitch vendors out of business. The Internet depression would also deny investments in new market entrants who would buy softswitch. Thus a whole market, VoIP long distance, is stifled in its development.

## Disruption in PBX Market

IP PBXs present some distinct advantages, providing the potential for significantly more convenience and flexibility than traditional telephone services. Specifically, they facilitate "unified messaging" with voice mail, e-mail, and faxes all in a single location. IP PBXs also make the "move/add/drop" process extremely easy (also known as a move/add/change or MAC and is purported to cost $500 each). Users can plug in anywhere within the corporation and have their regular telephone number. This allows new convergence-enabled features, combining voice, video, and data, and it lets corporations use a single architecture for all their traffic needs instead of maintaining separate networks.

IP PBXs and their related services are creating entirely new opportunities, with some companies targeting departments or offices within large corporations and others targeting small and medium businesses. IP PBXs are now poised to aggressively enter the mainstream market controlled by incumbent PBX providers such as Lucent, Nortel, and Alcatel. With their super-high reliability, high complexity, and long development cycles, traditional PBXs seem to overserve the marketplace, especially for small and medium businesses who cannot afford a PBX. Although the reliability of IP PBXs depends on the reliability of a corporation's underlying data network, IP PBXs' improvement trajectory does not seem to contain any fundamental barriers that would prevent them from getting good enough for the mainstream market. Small and medium businesses are clearly overserved by traditional PBX manufacturers and, in fact, many SMBs cannot afford most PBX offerings.

As IP PBXs improve, they present a very real disruptive threat to incumbent PBX manufacturers, who build many basic telecommunication services into their systems. IP PBXs are developing in a completely independent value network, making it difficult for incumbents to co-opt the technology. Additionally, IP PBXs entail different business models with different upgrade cycles, pricing structures, and feature deployments that are antithetical to the processes of incumbent providers. With their interdependent design (which is substantially less flexible than IP-based designs) and a long lead time to add features, traditional PBX manufacturers will find it difficult to adapt to the modular design inherent in IP-based systems.

A SIP-based IP PBX offers a great deal of simplicity relative to a conventional PBX. Potentially, an SIP-based PBX can be administered by an in-house IT administrator and does not require a factory representative to administer and maintain it the same way a conventional PBX does. Users can potentially provision their own services and set their own "follow me" schedules. As standards such as SIP gain market share, PBX systems will be ripe for disintegration, with new players competing at different points on the value chain.

Ironically, SIP will likely lead to a second wave of disruption, as IP PBX-like functionality migrates from hardware to software. Although these developments still remain within an enterprise, the internal transmission of voice as a data application finally creates a disruption to leap out of the enterprise and affect the incumbent service providers.[6] In summary, the IP PBX is cheaper, simpler, smaller, and more convenient to use than a conventional PBX.

# Conclusion

The disruption of Class 4 by softswitch is not absolute. A number of factors provide evidence both for and against the disruption of Class 4 in the converging market. Those factors, both pro and con, are semantic (sustaining versus disruptive technology), technical, and financial. The ultimate test of disruption, as addressed in Chapter 7, "Softswitch Is Just as Reliable as Class 4/5 Switches," is to determine which members of the value network have been disrupted. Different factors may cause different scales of disruption (if any). Also, the course of action of those potentially disrupted members of the value network will also determine if they are disrupted. Failure to identify a disruptive technology and take action along the guidelines set forth by Christensen can also determine the disruption or adoption of a disruptive technology.

---

[6]Mandel, Michael J. *The Coming Internet Depression*. New York, NY: Basic Books, 2000.

# Softswitch and Broadband

At the time of this writing, the vast majority of Internet subscribers world-wide access the Internet with a dial-up connection via a telephone line. This delivers a maximum bandwidth of 56 Kbps, which is adequate for email and surfing web sites. It is not adequate for video and other bandwidth-intensive applications. Some areas of the United States are served via a *digital subscriber line* (DSL) made possible by provisioning on the subscriber's telephone line where the service is usually a maximum of 256 Kbps. This is only marginally faster service, which is still not adequate for live video or the downloading of video on demand.

The chief bottleneck or limitation here is the *Public Switched Telephone Network* (PSTN), which was not designed to offer high-speed data services. *Incumbent Local Exchange Carriers* (ILECs) have tried to cobble together DSL/Class 5 configurations to serve their subscribers, but at the time of this writing no ILEC offers DSL service to more than 10 percent of their subscribers. True broadband at, say, 10 Mbps (no agreed-upon bandwidth exists for the term broadband) does not appear to be practical via the PSTN as we know it.

Cable TV companies offer broadband; however, cable TV companies have yet to roll out a truly converged data, video, and voice network. Where voice is offered with cable TV service, the cable TV company uses a Class 5 switch; that is, they must manage two networks, one cable TV and the other voice. No significant rollouts of converged voice, video, and data over the *Internet Protocol* (IP) via cable TV have taken place in the North American market. In order for a converged network offering voice, data, and video to come on the market, there must first be a strong consumer demand for it, and secondly there must be an easily deployed broadband technology.

# Converged Networks Independent of ILEC Infrastructure: "We Have the Technology"

It is possible that new technologies will enable a rapid and inexpensive rollout of residential broadband service. As we may recall from Chapter 2, "The Public Switched Telephone Network (PSTN)," and Chapter 3, "Softswitch Architecture or 'It's the Architecture, Stupid!'" a telephone network is three main elements: access, switching, and transport. The *Memorandum of Final Judgement* (MFJ) opened transport to competition in 1984. The bandwidth glut of recent years has made transport much more accessible, both in terms of engineering and cost.

The Telecommunications Act of 1996 was supposed to have opened the elements of switching and access to competition. At the time of this writing, only about eight percent of U.S. residential customers have a choice in local service providers. The failure of American regulatory agencies to enforce that law has left both elements largely closed to competition. However, as described in previous chapters, softswitch provides a technological bypass of the ILEC-owned Class 5 switch. This leaves access as the final element that must be opened to competition.

Access, also known as "the last mile," has been a sticky issue for would-be competitors to the ILECs. Failure to bypass this last mile is one of many reasons why the *Competitive Local Exchange Carrier* (CLEC) boom went bust in the late 1990s. The cost of stringing copper wires to each and every business and residence makes it economically impossible to compete with ILECs. Cable TV companies are presently the most prevalent alternative providers of wire-line service to the home. Cellular service providers only recently presented competition to wire-line services provided by an ILEC. Most cellular services use Class 4 and Class 5 switches in their networks.

# Access Alternatives to the PSTN and Cable TV

If softswitch provides a technological bypass of the RBOC Central Office, what good is that if a service provider must still contract with the RBOC for access to subscriber? Once access independent of the RBOC is achieved and voice is switched by a softswitch and transported via an IP backbone, incumbent telephone companies are then bypassed almost entirely. The following pages speculate on what alternative forms of access are available to a competitive service provider.

## Wi-Fi (802.11b Standard)

The broadband wireless Internet service is being built around a technology known as *wireless fidelity* (Wi-Fi), based on the *Institute of Electrical and Electronics Engineers'* (IEEEs) 802.11b standard. This is an increasingly popular networking standard that's used in wireless *local area networks* (LANs) in homes

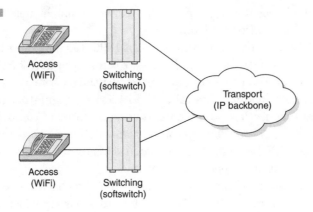

**Figure 13-1**
Components of a
network alternative
to the PSTN

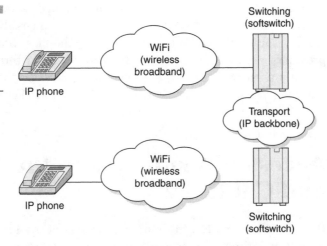

**Figure 13-2**
A more detailed
example of an
alternative to the
PSTN

and offices at speeds up to 11 Mbps, far faster than the peak 144 Kbps rate for
*third-generation* (3G) mobile-phone networks (see Figures 13-1 and 13-2).[1]

# Fiberless Optics

Another means of deploying broadband service rapidly without cables
and messy legal fights over rights of way is fiberless optics. Just as fiber

---

[1]Stone, Amey. "Wi-Fi: It's Fast, It's Here—and It Works." *Business Week*, April 1, 2002, www.
businessweek.com/technology/content/apr2002/tc2002041_1823.htm.

optics involves a laser beam shot over a fiber optic cable, fiberless optics, as the name would imply, shoots a laser beam through the air from a transmitter to the subscriber's receiver. To date, fiberless optics is a popular alternative to fiber optic cable in dense urban business settings, but it has yet to be deployed in residential applications in any significant numbers.

# $500 Billion Economic Benefit of Converged Broadband Networks

A wave of opportunity for softswitch applications is in the making. Most of it lies in the form of broadband deployment. The demand for softswitch solutions lies largely with the ubiquitous availability of broadband. In their April 2001 white paper, "The $500 Billion Opportunity: The Potential Economic Benefit of Widespread Diffusion of Broadband Internet Access," Robert Crandall and Charles Jackson point to an economic benefit of $500 billion per year for the American economy if broadband Internet access were to be as ubiquitous as land-line phones.

As the uses of broadband multiply, the value to subscribers rises far above the monthly subscription price. This is the *consumer surplus* from the innovation. Producers of new services that rely on broadband (i-Mode-type services, Net2Phone, and so on), creators of products used in conjunction with broadband service (softswitches, media gateways, IP phones, residential gateways), and even producers of the broadband service itself also gain from the greater diffusion of broadband. The *producer surplus* that is generated by sales is a real benefit to producers and therefore to the economy. At present, no more than 8 percent of American households subscribe to a broadband service, only slightly more than 50 percent subscribe to an Internet service of any kind, and 94 percent subscribe to ordinary telephone service.[2] Were broadband to become ubiquitous, it would resemble current telephone service in its household penetration.

---

[2] The number of broadband subscribers (DSL plus cable modems) was 7.3 million as of March 2001. See Hammond. "Failure of Free ISPs Triggers First-Ever Dip, To 68.4 Million Online Users: Cable Modem Boom Continues, As DSL Sign-ups Lag," *Telecommunications Reports*, April 2001. The estimates for Internet and telephone service are from the author's tabulations using the Current Population Survey for August 2000.

# Broadband Access and Telephone Services

The delivery of broadband services requires the use of existing telecommunications networks (initially) plus substantial expenditures on new electronics within such networks (multiplexers, routers, and so on) as well as expenditures on consumer access devices. Therefore, any increase in the demand for broadband will generate additional revenues not only for the suppliers of that service, but also for the suppliers of the equipment used to deliver the service.

For broadband services, Crandall and Jackson conservatively estimate that 50 percent of the revenues are spent on the electronics and related equipment to deliver the service. They also estimate that 20 percent of the remainder accrues as "quasi-rents" to the supplier of the communications service itself. Under these assumptions, Crandall and Jackson may calculate the producers' surplus for an expansion of broadband to 50 percent of households and to "universal service" proportions of 94 percent of households. The results range from $6.2 billion to $17.6 billion per year.

Soon the demand for broadband will reflect not only the growing potential uses of the Internet, but also the prospect for using these broadband connections to obtain voice telephone services currently provided over a narrowband connection. The use of broadband access to carry voice (ordinary telephone calls) as well as data will deliver substantial savings to consumers that are not captured in current demand estimates.

A variety of options for *Voice over IP* (VoIP) exist in the residence or small business. An IP handset connects directly to the IP network via an Ethernet connection (a router or hub). Media gateways for the residence or small business range from one to eight ports that connect analog handsets to the IP network via a standard RJ-11 telephone jack. The PC, loaded with the appropriate software, can also be used with the speakers and microphone resident on the sound card in the PC. For those who find using a PC to be an anthropological challenge, an analog handset can be attached to the PC with all the feel of a conventional telephone.

A broadband connection can support several voice connections. Using uncompressed VoIP at 64 Kbps, hundreds of simultaneous voice conversations can occur simultaneously over a 10 Mbps broadband IP connection. The current structure with two networks in the home (a voice and an IP network) and two connections to the outside world (a narrowband analog connection and a high-speed digital connection) is inefficient. However, the transition to Internet telephony will take a few years. The existing inside wiring, designed for voice systems, will probably be replaced by Wi-Fi applications. The analog telephones must be replaced or connected to digital adapters. As addressed in Chapter 5, "SIP: Alternative Softswitch Architecture?" *Session Initiation Protocol* (SIP) and ENUM, a means of mapping a telephone number from an email address or URL or vice versa, provide for the mapping of phone numbers to email addresses. The Telecommunications Act of 1996 provides for local number portability.

The cost savings from integrated access will be significant. IP telephony via broadband eliminates the need for second or third lines in households. The *Federal Communications Commission* (FCC) has estimated that the average household spent $55 per month on local and long-distance telephone service in 1999, and 0.289 additional lines existed for each household with telephone service.[3]

Within a few years, broadband access will permit consumers to substitute other services for the ones that now cost $55 per month. The FCC estimates that the average residence spends $34 per month for local telephone service and $21 for long-distance telephone service. Part of that local telephone service cost is for the loop used for the broadband service. Consumers continue to incur most of those loop costs when broadband service is used, but they avoid the cost of the analog line card, the voice switch, and the voice transmission lines.

IP telephony should lower the costs of both local and long-distance telephone service while providing residences with the equivalent of several telephone lines. Crandall and Jackson estimate that such savings could average $25 per month per household. In addition, households with broadband service would get the equivalent of multiple voice telephone lines. We estimate that this additional service or option of service could be worth $10 per month to the average household.[4] Thus, in the long run, broadband access could deliver voice communication benefits of about $35 per month, or $420 per year, to the average household with telephone service. If we assume that 122.2 million households have telephone service, these benefits would total $51.4 billion per year, assuming no growth in voice usage. The actual value could be much higher.

---

[3]Federal Communications Commission. "Trends in Telephone Service." December 21, 2000. www.fcc.gov/Bureaus/Common_Carrier/Reports/FCC-State_Link/IAD/trend200.pdf.

[4]The FCC's numbers indicate that the average household with telephone service has 1.289 access lines and pays local service fees of $34 per month. Assuming that all lines cost the same (which is not quite right but is reasonable), the average household with telephone service in 1999 paid $7.62 per month for additional line service. If those households without a second line today place an average value of no more than $3.40 per month for a second line of service, then the average household will value a second line at $10 per month or more.

The substantial economic benefits (principally savings from expenditures on telephone service) created by providing multiple services over a high-speed line almost cover the cost of a high-speed line. It is estimated that the benefits of $35 per month are created by a broadband connection that costs $40 per month. These savings are one reason why it is reasonable to expect that the fraction of households with high-speed access services will ultimately approach the number that has telephone service today (see Table 13-1).[5]

These two different approaches provide quite comparable estimates of the prospective consumer benefits from broadband. From these two approaches, we conclude that the annual consumer benefits from broadband could eventually reach $300 billion per year. In addition, producers will also benefit from an increased demand for electronic equipment used in the delivery of broadband service, increased spending on household computer and networking equipment as well as on household entertainment, and improvements in health care delivery. These benefits could easily amount to another $100 billion per year if broadband becomes ubiquitous. Thus, a reasonable figure for the total annual benefits to the U.S. economy due to the widespread adoption of broadband access in all its forms (DSL, cable modems, satellites, 3G wireless, and others) is $400 billion per year.

A faster rollout of high-speed access services gives us these benefits earlier. Under optimistic but still reasonable scenarios, the net present value of a faster rollout of high-speed access could be as high as $700 billion, and a midrange estimate of the value of a faster rollout is $500 billion.

**Table 13-1**

One of many possible scenarios comparing broadband with unconverged telecommunications services

| Service | With broadband | Without broadband |
|---------|----------------|-------------------|
| Telephone (one line) | $40 | $34 |
| Internet service provider (ISP) | $25 | $25 |
| Cable TV | Potentially included | $50 |
| Long distance | Potentially included | $100 |
| Totals per month | $65 | $209 |

[5]Convergence in the form of residential broadband can be more economical for the service provider. Why have two line cards at the telephone central office, one in the voice switch, and one in the DSLAM?

# National Defense Residential Broadband Network (NDRBN)

Just as the interstate highway system in this country was introduced as the National Defense Highway System and initially funded as a defense project, a comparable program could provide a majority of American households with broadband Internet access and a multitude of societal and economic benefits. Also, many benefits, both military and civilian, would be provided by a *National Defense Residential Broadband Network* (NDRBN). The military applications would include the following:

- Service men and women would have a means of working at remote sites, including their residences, which could save the *Department of Defense* (DoD) millions of dollars annually in telephone, travel, and relocation expenses.

- Military reservists could stay in touch with their units via email, high-bandwidth web sites, video conferencing, long-distance calling, and video on demand. Military veterans subject to recall could also stay informed of military issues.

- In exchange for subsidized broadband Internet access, civilian computers could be networked to DoD computers to perform distributed computing projects (crack enemy codes, map human genomes, and so on).

- Law enforcement agencies could better share terrorist alerts with federal and military intelligence sources via email, high-bandwidth web sites, free long-distance telephony, video conferencing, and video on demand.

Civilian applications and benefits would include the following:

- The U.S. dependency on foreign oil would be reduced by empowering Americans to telecommute using video conferencing and other communication technologies. This applies equally to reducing costly business travel utilizing the same technologies.

- Work forces could be dispersed to less dense population areas, presenting less desirable targets for terrorists (compare rural Colorado to Lower Manhattan). This could be an engine for rural economic development nationwide.

■ Education could be enhanced over great distances for both adults and school-age children via email, high-bandwidth web sites, video conferencing, and video on demand.

■ Worker productivity could be boosted by jump-starting the now beleaguered *information technologies* (IT) sector of our economy.

The NDRBN would use multiple technologies (DSL, cable TV, and wireless) to deliver broadband services to the last mile. This would ensure redundancy and survivability in the network and encourage investments in a wide range of technologies and businesses creating millions of jobs. Simply planning this network could provide job security for thousands of Americans displaced by the "telecom meltdown."

At present, no more than 8 percent of households subscribe to a broadband service, only slightly more than 50 percent subscribe to an Internet service of any kind, and 94 percent subscribe to ordinary telephone service. Were broadband to become ubiquitous, it would resemble current telephone service in its household penetration and resulting social and economic benefits.

# Better Living Through Telecommunications: The Social Rewards of Softswitch

The "big so what?!" of any technology is measured by its impact on society (positive or negative). Softswitch, in conjunction with broadband internet access, potentially offers a number of positive impacts on society including the negation of commuting, affordable housing, and the improvement of family life.

## Essay One—If It Hurts to Commute, Then Don't Commute

No traffic problems exist in the industrialized world. A large percentage of the drivers on the developed world's highways are driving to offices where they work on computers and telephones the majority of their work day. Are we to believe those in the Audis and Land Rovers do not have telephones and computers at home? Why do they clog highways to go somewhere to do something they could just as well do in their homes? Why are tax dollars consumed by their demand for highways, parking garages, and other wasteful forms of transportation infrastructure that only serve to breed more congestion?

The primary direct benefits of telecommuting occur from the reduction in travel required by the employee and the reduction in infrastructure costs at the office. But a significant secondary benefit is a reduction in congestion costs. The Texas Transportation Institute's Urban Mobility Study report estimates the costs of traffic congestion in 68 urban areas.[6] Its 2001 report states

> Congestion costs can be expressed in a lot of different factors, but they are all increasing. The total congestion "bill" for the 68 areas in 1999 came to $78 billion, which was the value of 4.5 billion hours of delay and 6.8 billion gallons of excess fuel consumed. To keep congestion from growing between 1998 and 1999 would have required 1,800 new lane-miles of freeway and 2,500 new lane-miles of streets—or—6.1 million new trips taken by either carpool or transit, or perhaps satisfied by some electronic means—or—some combination of these actions. These events did not happen, and congestion increased.

Analysis of the TTI report shows that 80 percent of these $78 billion in costs occurs in only 24 cities (comprising most of the larger cities in the United States), and 90 percent occurs in 36 cities. In Los Angeles, traffic congestion imposes estimated costs of $1,000 per person per year.

Various studies estimate that 20 to 40 percent of jobs permit telecommuting at least part of the time. If we assume that 30 percent of the jobs permit telecommuting an average of 20 percent of the time (1 day per week or 50 days per year) and that the average commuter trip for a telecommuter is 20 minutes, we can calculate the potential savings in travel costs and congestion. The savings in travel time is as follows:

$$(180 \text{ million civilian labor force}) \times (30\% \text{ possible telecommuters}) \times (33 \text{ hours/year}) \times (\$6.20/\text{hour}) = \$11.1 \text{ billion per year}$$

Similarly, the savings in travel costs are as follows:

$$(180 \text{ million civilian labor force}) \times (30\% \text{ possible telecommuters}) \times (450 \text{ miles/year}) \times (\$0.3/\text{mile}) = \$7.3 \text{ billion per year}$$

---

[6]The hourly value of $12.40 is used in the congestion analysis by the Texas Transportation Institute in its urban traffic studies discussed here (Texas Transportation Institute. "The 2001 Urban Mobility Report." College Station, TX: The Texas A&M University System, May 2001, http://mobility.tamu.edu).

Summing up, the potential savings from telecommuting is $11.1 billion per year travel time for telecommuters, $7.3 billion per year for travel costs, and $4.7 billion per year for reduced (external) congestion costs, yielding a total savings of over $23 billion per year. Note that this is not an estimate of the savings from the accelerated deployment of broadband access; this is an estimate of the total transportation system savings from the widespread adoption of telecommuting. Assuming that these savings grow at a rate similar to the general growth rate we assume for the economy, these savings could be as much as $30 billion in 10 years.

Why do commuters persist following an Industrial Age habit in the Information Age? Sadly, we have translated the drill press as the computer. Instead of a factory, we now have paperwork factories in the form of downtown office buildings and outlying office parks. If our white-collar work force could only break out of their Industrial Age lockstep mentality and work at their homes or local teleworking centers, a meaningful percentage of our traffic would simply evaporate.

One objection to this solution for traffic congestion solutions is that white-collar workers need face-to-face contact to coordinate their work. True, but does it have to be exactly at 8 A.M. through 5 P.M. Monday through Friday? Wouldn't attendance at biweekly staff meetings suffice, the rest of the time being spent at home offices or teleworking centers nearer to home?

Another objection is the need for socialization. First, one would speculate that if one's only friends are those in the office, that individual should work at "getting a life." Secondly, does that socialization have to be all day Monday through Friday? We can take our cue from America's first home workers: farmers. Much of rural day-to-day socialization evolves around a visit to a neighbor for coffee or a stop in the local café for the same. This practice could be adapted in suburbia.

A more concrete objection, especially among high-tech workers, is the need for expensive computers with high-speed connections to the Internet. Many corporations who have weighed the cost per square foot of maintaining work space in downtown offices or out at the local tech center have found it cheaper to send their people home with an expensive computer and offer the employee a stipend for his or her home office. Competition to offer high-speed Internet access to the home by phone line, cable TV, or wireless would negate this Internet bandwidth objection.

So is there a high-tech, expensive solution for our so-called traffic problem that elected leaders can plan and impose over the next few decades? The best thing elected leaders can do is *nothing*. Once traffic becomes enough of a painful experience, those that don't *have* to brave traffic morning and night simply will not do so. They will stay home to do their work.

They will slowly realize that the Information Age is driven by ideas. The formation of ideas is not limited to 9 to 5 Monday through Friday. Nor are ideas limited to one place geographically, especially not a corporate cubicle. Perhaps we will ask that question from World War II gas-rationing pitches, "Is this trip really necessary?"

Concrete roads were first developed by the Romans to move their chief commodity, labor (soldiers and slaves), quickly and efficiently throughout the empire. In the Information Age, our chief commodity is ideas. Ideas don't need roads for their transportation. Ideas move well over fiber optic cable, phone lines, coaxial cable TV cables, or even the airwaves. We must ask our leaders to compare the cost per mile of pavement (millions of dollars) to fiber optic cable (approximately $25,000) or even wireless broadband. Its time to tell elected officials that they need to plan for the 2020s (the Information Superhighway) not the 1920s (the Lincoln Highway).

## Essay Two: Affordable Housing Is Where You Find It

A societal ill dictates that good-paying jobs are only found in downtown high rises or in outlying office parks. Because that is "where the money is," white-collar or Information Age workers strive to live within an easy commute of those offices. This in turn breeds congestion as everyone attempts to spend exactly the hours of 9 to 5 Monday through Friday in those offices.

Most government officials have not caught on to this yet, but new telecommunications technologies have largely negated the need for white-collar or Information Age workers to drive to an office. In the bad old days, there were factories. The machinery in a factory cost a lot of money. Workers had to go to the factories because that was where the jobs were. With the coming of computers and the Internet, white-collar and Information Age workers can work from their homes in the suburbs. As telecommunications improve in more rural areas, white-collar and Information Age workers can do their jobs from much farther afield and improve their net worth.

The average sale price of a single-family home in Denver, Colorado, at the time of this writing (Spring 2002) is over $270,000. Sale prices of single-family homes in small-town Iowa, by contrast, hover at $30,000. Imagine a white-collar or Information Age worker making his or her big city income in a low-cost area like a small town in Iowa. The savings in mortgage payments are obvious. If one went over his or her monthly budget, other significant savings could be found in such a lifestyle change (telecommuters don't rack up as many miles on their cars). By saving as much as possible

of that big city income while living in a low-cost area, financial independence could be realized in a short period of time. Contrast that with the
bumper sticker, "I owe, I owe, it's off to work I go," representing the con-
sumptive/dependent lifestyle of the suburbanite.

Imagine a society where we didn't have to move every few years to stay
competitive in our work. Imagine being able to stay in one place most of our
working lives as our work was not dependent on relocating to another city
to take another job with the subsequent loss of contact with friends and
family. Information Age work is not dependent on working from a specific
cubicle in a specific office building. It is dependent on having a consistent
marketable skill. In theory, an Information Age worker should be able to
change jobs or contracts repeatedly over a working life and never have to
change residences. Imagine successive generations living under one roof—
no mortgage, no moving expenses, and no closing fees.

## Essay Three: Family Values

Why are there latch-key children? The answer is simple; it's because some
people think that the only place a job can be done is in a specific office at a
specific time. Therefore, Mom and Dad must be at those specific places at
those specific times, which leaves their children coming home from school to
an empty house and, perhaps, trouble.

If it were better understood that most white-collar jobs could be performed from a home office, then Moms and Dads would not need to be
absent when their children arrived home from school. Hence, social ills
associated with absent parents would be greatly diminished.

# The Role of Softswitch in Better Living Through Telecommunications

At about 1920, half the American population either lived on farms or in
farming communities. Today that figure is less than five percent. What
happened over the last 80 years? Obviously the farming half of the population moved to the cities in search of better pay. Cities have always been
communication hubs (railroad, highways, telephones, and the Internet).
Also, there has historically been a perception that cities have more enter-

tainment (theater, opera, live music, and cinema) than what could be found in the country.

The telecommunications revolution can make both of those perceptions obsolete. Historically, rural areas are the last to get the newest telecommunications technologies because they pose a lower rate of return for the service provider. Much of rural America's telephone and electric infrastructure was built by cooperatives of farmers pooling their labor and money to install telephone and electric poles and wires. This scarcity of a telecommunications infrastructure extends to the suburbs of the largest cities in the United States. As of 2002, only an estimated eight percent of U.S. households had access to broadband services. This is due largely to a lack of infrastructure for broadband and perhaps a reluctance on the part of incumbent service providers to make the investment on DSL infrastructure when they could be retaining the cash to make their stock more attractive or pay down mountains of debt.

Softswitch, by virtue of being simpler, smaller, cheaper, and more convenient for competitive service providers to deploy, enables more service providers (ISPs, application service providers, power companies, municipalities, cable TV companies, wireless service providers) to reach more customers. Most importantly, this would allow data service providers to offer voice services, which offer higher margins than data services. The prospect of generating revenue from both data and voice should motivate service providers to enter as many markets as possible. This trend could be enhanced when service providers can bypass the facilities of incumbent service providers. Even if alternative service providers don't specifically charge for voice, bundling it with data could prove attractive.

Softswitch is a technology that can ignite competition for telecommunications services. What little competition for telecommunications services the Telecom Act of 1996 inspired was for business services in business districts throughout the United States in the late 1990s. Competitive service providers sought to "cherry pick," that is, offer service only to business subscribers in high-density areas, thus maximizing the return on investment in infrastructure in city centers. This ultimately led to hyper-competition, a "race to the bottom" in pricing and ultimately the shakeout and bankruptcy of many service providers. This hyper-competition should have the effect of driving service providers to find less competitive markets (suburbs and rural areas) where they would enjoy less volatility. In some metropolitan markets, the price of a data T1 has dropped to $300 per month (see Table 13-2).

The $300 per month for 1.54 Mbps of data services would roughly equate to what an average suburban family now pays for telecommunications services per month. Using the latest in IP technologies, the service provider

**Table 13-2**

Tabulation of residential telecommunications expenditures per month

| Service | Monthly cost |
| --- | --- |
| Telephone (two lines) | $75 |
| Internet service provider | $25 |
| Broadband service (DSL) | $50 |
| Cable TV | $50 |
| Long distance | $100 |
| Total | $300 |

can offer all the services listed in Table 13-2. Therefore, a residential customer is arguably just as lucrative as a small business customer.

Once service providers can be coaxed into offering broadband, softswitched services in suburban and rural markets, more residents can work from home. If no service provider will rise to the occasion, there is little reason why potential subscribers cannot take action the way their grandparents did. By establishing the white-collar equivalent of the farmer's co-op telephone company by banding together and pooling their resources to form co-ops, subscribers could buy and install equipment necessary to deliver broadband IP services.

# Conclusion

This chapter explored more "big so whats" related to the impact of softswitch technologies on society. New technologies are now available that, when combined with softswitch solutions, potentially replace the PSTN. Use of Crandall and Jackson's economic analysis of broadband technologies provides a dollar and cents metric for the impact on the American economy if broadband and, tangentially, softswitch technologies were to be as widely deployed as residential telephone service.

Softswitch, when coupled with broadband IP networks, will enable an improved standard of living in the form of telecommuting, lower real estate prices, and improved family lives. If the private sector cannot build such a network, it may be necessary for governments, local or national, to build such a network just as governments worldwide have built other "lines of communications" in the form of roads and railroads for a well-demonstrated common good.

# Past, Present, and Future of Softswitch

# History of Softswitch

What is the origin of the term and technology of softswitch? The term *softswitch* was coined by Ike Elliott when he worked for MCI to develop an interface for *Interactive Voice Response* (IVR) with a circuit switch. In 1997, Mr. Elliott left MCI to join a new startup, Level3 Communications, a pioneering all-*Internet Protocol* (IP) backbone communications company.

There he teamed up with Andrew Dugan and Mauricio Arongo, a team that went on to coin the terms *call agent* and *media gateway*. At this time, no major telecom vendor offered technology referred to as a *media gateway controller* (MGC), which would be crucial to Level3's plan to roll out an all-IP backbone. Decision makers at Level3 decided early on in their history to develop their own softswitching technology rather than wait for telecom vendors to offer technology compatible with their goals. In early 1998, Level3 began discussions with a company called Xcom, which had developed technology for managed modem service. This technology enabled a softswitch specification for a message interface between a softswitch and a modem bank, which is what *Internet service provider* (ISP) subscribers dial into to access the Internet. Xcom's specification was a device control protocol. In April 1998, Level3 purchased Xcom for $165 million in stock. After that, Level3 renamed their newly acquired protocol the *Internet Protocol Device Control* (IPDC) and founded the IPDC *Technical Advisory Council* (TAC).

In September 1998, Level3 teamed with Christian Huitema of Bellcore and author of the *Signaling Gateway Control Protocol* (SGCP) to draft the specification for the *Media Gateway Control Protocol* (MGCP) for the *Internet Engineering Task Force* (IETF). Until this time, the Level3 softswitch team was developing a platform that would manage modems and did not focus on softswitch solutions for *Voice over IP* (VoIP). In the Spring of 1999, Level3 entered negotiations with Lucent Technologies in the search for a softswitch that would switch VoIP. In the winter of 2000, Level3 ended its evaluation of Lucent Technologies' softswitch and reopened their product evaluation to find a softswitch to meet their needs. Level3 eventually selected Sonus Networks' softswitch. In addition to the Sonus Networks' softswitch, Level3 continued to develop their own softswitch in house known as the Viper (Voice over IP Enhanced Routing). As of this writing, some 13 billion minutes of VoIP is utilized per month. Level3 requires only 50 people to maintain and operate this part of their network.[1]

---

[1]Interview with Ike Elliott, Senior Vice President, Global Softswitch Services, Level3 Communications, May and June, 2002.

# The Present of Softswitch: Case Studies

One issue with new technologies is the preponderance of hype, that is, products or technologies that promise great things but deliver no real-world applications. The purpose of the next few pages is to document, without endorsing any one product or vendor, real deployments of softswitch in both the Class 4 and 5 space where real softswitches handle real traffic and earn real money as a result.

## Class 4 Replacing the Class 4 Switch in Long-Distance Applications: Fusion, Sonus Networks, and NexVerse

Softswitch vendor Sonus Networks offers a success story in their deployment with Fusion, a Japanese service provider. Fusion wanted to build a packet-based network, rather than a circuit-switched one, to take advantage of the inherent cost savings and flexibility. Fusion noted that in the past when telecommunication carriers created separate networks for voice, data, Internet, and other services, their cost structures increased because they lost flexibility and economies of scale. Fusion wanted an end-to-end converged solution for voice and data communications.

To achieve its goals, Fusion needed to deploy a carrier-class all-IP network quickly. The company wanted to gain a first-mover advantage as the only alternative long-distance carrier in Japan's newly deregulated telecommunications market. To make its next-generation carrier business model work, Fusion had to quickly launch a nationwide network that would enable it to generate increasing amounts of revenue while simultaneously keeping costs down.

Fusion launched its next-generation VoIP network with Sonus Networks' GSX9000™ Open Services Switch, which performs the media gateway function defined in the *International Softswitch Consortium* (ISC) reference architecture (described later in this chapter). The GSX9000 employs a distributed architecture and is NEBS Level 3 certified and fully redundant for deployment in a variety of configurations in either central offices or *points of presence* (POPs), such as colocation hotels, for example. The Sonus Insignus™ Softswitch splits ISC-defined softswitch functionality into a set of comprehensive modular software components that can be configured in either a distributed or centralized fashion to accommodate a wide range of network topologies. The solution uses both the PSX Policy Server and the SGX *Signaling System 7* (SS7) Gateway modules for the Insignus Softswitch. Fusion selected Sonus based on the high density and reliability of the offering. The Sonus solution helped Fusion seize a market opportunity that would be impossible without next-generation packet voice technology.

Fusion went live with its network in April 2001, rolling out its low-cost, fixed-rate service to some 360,000 subscribers throughout Japan. In a matter of months, the number of subscribers exceeded 1,100,000 and continues to grow rapidly. Japanese versions of the Sonus solution were installed in 18 *network operations centers* (NOCs) distributed throughout Japan. In each of these NOCs, the GSX9000 switches processed many thousands of calls simultaneously, delivering high-quality voice and the scalability necessary to support millions of subscribers.

The Sonus PSX Policy Server module in the Insignus Softswitch establishes a solid service development environment that provides advanced service selection and routing capabilities. Additionally, the Sonus SGX SS7 Gateway modules operate in nine switching offices, providing seamless interconnection to the Japanese PSTN. Fusion plans to expand the services to more business customers, beginning with a new *primary rate interface* (PRI) direct connection service for *private branch exchange* (PBXs). Fusion also plans to deliver other new services, such as virtual call centers and toll-free calling, which the company would not be able to offer without this next-generation network.[2]

## Replacing the Class 5 Switch: NorVergence and MetaSwitch

At the time of this writing, the consensus among the telecommunications cognoscenti is that softswitch as a Class 4 replacement is a no-brainer, given that it deals largely with long-distance transport via IP. Softswitch as a Class 5 replacement in the context of a port-for-port or feature-for-feature substitute is not quite there.

---

[2]International Softswitch Consortium. "Case Study: Sonus Networks Provides a Next-Generation Voice/Data Infrastructure for Fusion Communications." A white paper available at www.sonusnetworks.com.

However, MetaSwitch, a vendor of Class 5 replacement softswitch solutions, begs to differ. MetaSwitch recently completed the installation of their Class 5 replacement solution with NorVergence, an emerging managed care/total care application provider. MetaSwitch installed its VP3000 next-generation Class 5 switch in NorVergence's New Jersey operations center. The switch delivers packet-based voice and data services to NorVergence's small and medium-sized business customers.

NorVergence's goals for the project were to take advantage of the latest broadband packetized data and voice technologies to deliver high-quality services over data circuits in the most cost effective way possible. The initial deployment involved a single VP3000 switch transporting packetized voice over data networks to major *interexchange carriers* (IXCs) for termination to the PSTN.

NorVergence guarantees its customers' telecommunications costs will be cut by at least 30 percent. NorVergence calls this the *Merged Access Transport Intelligent Xchange* or the MATRIX™ system, an alternative to the circuit-switched model. The MetaSwitch VP3000 transports toll-quality packetized voice with 99.999 percent reliability, achieving cost savings through integrated management and the delivery of regional, long-distance, and Internet access services over a single broadband local loop.[3]

# The Future of Softswitch: ISC Reference Architecture and the ISC

The ISC is an industry organization dedicated to the promotion of the softswitch industry primarily in the form of promoting standards and the interoperability of softswitch technologies. Its 120-plus members meet biannually to promote their industry. The ISC has a number of working groups that debate and promote features, marketing, and protocols (see Table 14-1). The ISC organizes interoperability test events, holds educational conferences, and establishes industry works groups to address issues vital to the industry.

---

[3]Interview with Tony Downes, Vice President of Marketing, June 5, 2002.

**Table 14-1**

Working groups of
the ISC

| Working Group | Focus |
|---|---|
| Legal Intercept Working Group | Coordinates work between members and legal and regulatory agencies (such as between the FCC and the FBI). |
| Applications Working Group | Promotes a faster introduction of enhanced services that combine voice, Internet, messaging, and other content. |
| Device Control Working Group | Covers the interoperability of independently developed softswitch devices through a single, well-recognized standard. |
| Network Boundary Functionalities | Documents and reviews requirements from carriers and enterprise markets for VoIP applications. |
| Marketing Working Group | States the ISC's goals and facilitates the acceptance and implementation of the ISC's proposed architecture and protocols. |
| SIP Working Group | Addresses the collaboration of multiple softswitches during the setup and teardown of calls using the *Session Initiation Protocol* (SIP). |

Source: ISC

The ISC is not a standards body. It promotes standards through interoperability test events, specifications, reference implementations, and development resources for companies and individuals seeking to develop applications based on the standards set by the IETF and the *International Telecommunications Union* (ITU), which are standards bodies.

The following pages contain the ISC's reference architecture. This presents a fitting ending for this book as the reader has been informed about the building blocks of softswitch architecture, VoIP, how softswitch overcomes objections of the past, and the economics and regulatory issues of softswitch. The reader is now properly prepared to digest the reference architecture for softswitch as envisioned by the ISC.

One objection to VoIP and softswitch is that some incumbent service providers consider this technology to be too "new;" that is, it lacks standards and an industry-wide consensus on architecture, standards, and technology. The following pages from the ISC should go a long way in establishing a framework for service providers regarding protocols and components for their new networks.

As with any industry in its infancy, the softswitch business desires standardization, particularly in regards to terminology. The rest of the chapter except for the concluding section is excerpted from the ISC document "Reference Architecture."[4] The purpose of this document is to standardize first and foremost the terminology used in describing components of softswitch architecture. It also seeks to promote open network architectures and standard interfaces. The document also provides a guideline for vendors and service providers alike for standardizing softswitch architecture.

# Functional Planes

The functional planes represent the broadest level of separation between the functional entities in VoIP network. There are four distinct functional planes employed by the ISC to describe the functioning of an end-to-end VoIP network: Transport, Call Control & Signaling, Service & Application, and Management (see Figure 14-1).

**Figure 14-1**
Functional planes for
ISC Reference
Architecture
(Source: ISC)

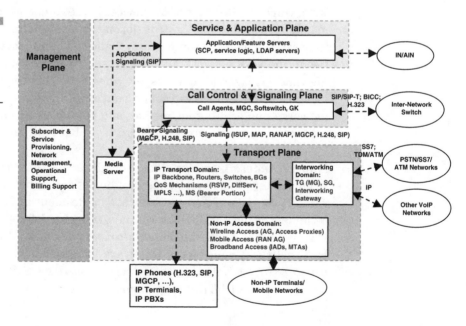

---

# Transport Plane

The Transport Plane is responsible for the transport of messages across the VoIP network. These messages may be for call signaling, call and media setup, or media. The underlying transport mechanism(s) for these messages may be based on any technology that satisfies the requirements for carrying these types of traffic.

The Transport Plane also provides access for signaling and media with external networks, or terminals to VoIP networks. Often the Transport Plane devices and functions are controlled by functions in the Call Control & Signaling Plane.

The Transport Plane is further divided into three domains: IP Transport Domain, Interworking Domain, and Non-IP Access Domain.

**IP Transport Domain**    The IP Transport Domain provides the transport backbone and routing/switching fabric for transporting packets across the VoIP network. Devices like routers and switches belong to this domain. Devices that provide *Quality of Service* (QoS) mechanisms and policies for the transport also belong to this domain.

**Interworking Domain**    The devices in the Interworking Domain are primarily responsible for the transformation of signaling or media received from external networks into a form that can be sent among the various entities in the VoIP network and vice versa. It consists of devices like Signaling Gateways (signaling transport conversion between different transport layers), Media Gateways (media conversion between different transport networks and/or different media), and Interworking Gateways (signaling interworking on the same transport layer but with different protocols).

**Non-IP Access Domain**    The Non-IP Access Domain applies primarily to non-IP terminals and wireless radio networks that access the VoIP network. It consists of Access Gateways or Residential Gateways for non-IP terminals or phones, ISDN terminals, *Integrated Access Devices* (IADs) for DSL networks, Cable Modem/*Multimedia Terminal Adapters* (MTAs) for HFC networks, and Media Gateways for a GSM/3G mobile *radio access network* (RAN). Note that the IP terminals, like a SIP phone, will directly connect to the IP Transport Domain, without going through an Access Gateway.

## Call Control & Signaling Plane

The Call Control & Signaling Plane controls the major elements of the VoIP network, especially in the Transport Plane. The devices and functions in this plane carry out call control based on signaling messages received from the Transport Plane, and handle establishment and teardown of media connections across the VoIP network by controlling components in the Transport Plane. The Call Control & Signaling Plane consists of devices like the Media Gateway Controller (a.k.a. Call Agent or Call Controller), Gatekeepers and LDAP servers.

## Service & Application Plane

The Service & Application Plane provides the control, logic and execution of one or more services or applications in a VoIP network. The devices in this plane control the flow of a call based on the service execution logic. They achieve this by communication with devices in the Call Control & Signaling Plane. The Service & Application Plane consists of devices like Application Servers and Feature Servers. The Service & Application Plane may also control specialized bearer components, such as Media Servers, that perform functions like conferencing, IVR, tone processing, and so on.

## Management Plane

The Management Plane is where functions such as subscriber and service provisioning, operational support, billing and other network management tasks are handled. The Management Plane can interact with any or all of the other three places through industry standard (for example, SNMP) or proprietary protocols and APIs.

# Functional Entities

The functional entities are the logical entities of a VoIP network. This section describes the major functional components of the ISC Reference

Architecture. Note that these are functions and *not* physical product descriptions. The various functions may physically reside on standalone devices or in various combinations on multifunction platforms. As such, there are virtually an unlimited number of ways to bundle the various functions into physical entities.

The diagram (in Figure 14-2) shows 12 different functions. Understanding the autonomy of all 12 functions is an important characteristic of the ISC Reference Architecture.

## Media Gateway Controller Function (MGC-F) a.k.a. Call Agent or Call Controller

The MGC-F provides the call state machine for endpoints. Its primary role is to provide the call logic and call control signaling for one or more media gateways.

MGC-F characteristics are as follows:

- Maintains call state for every call on a Media Gateway
- May maintain bearer states for bearer interfaces on the MG-F
- Communicates bearer messages between two MG-Fs, as well as with IP phones or terminals

**Figure 14-2**
Functional entities of ISC Reference Architecture (Source: ISC)

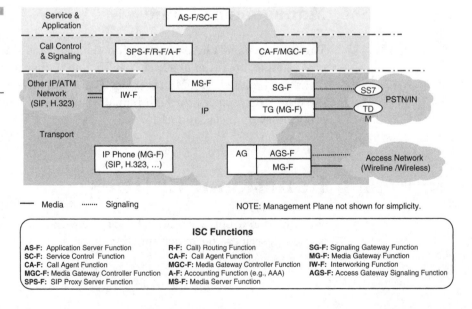

—— Media ······· Signaling          NOTE: Management Plane not shown for simplicity.

**ISC Functions**

| | | |
|---|---|---|
| **AS-F:** Application Server Function | **R-F:** Call) Routing Function | **SG-F:** Signaling Gateway Function |
| **SC-F:** Service Control Function | **CA-F:** Call Agent Function | **MG-F:** Media Gateway Function |
| **CA-F:** Call Agent Function | **MGC-F:** Media Gateway Controller Function | **IW-F:** Interworking Function |
| **MGC-F:** Media Gateway Controller Function | **A-F:** Accounting Function (e.g., AAA) | **AGS-F:** Access Gateway Signaling Function |
| **SPS-F:** SIP Proxy Server Function | **MS-F:** Media Server Function | |

- Acts as conduit for media parameter negotiation
- Originates/terminates signaling messages from endpoints, other MGC-Fs and external networks
- May interact with the AS-F for the purposes of providing a service or feature to the user
- May manage some network resources (for example, MG-F ports, bandwidth and so on)
- May provide policy functions for endpoints
- Interfaces to R-F/A-F for call routing, authentication, and accounting
- May participate in management tasks in a mobile environment (mobility management is generally part of the CA-F)
- Applicable protocols include H.248 and MGCP

**Call Agent Function (CA-F) and Interworking Function (IW-F)**
The CA-F and IW-F are subsets of the MGC-F. The CA-F exists when the MGC-F handles call control and call state maintenance. Examples of CA-F protocols and APIs include:

- SIP, SIP-T, BICC, H.323, Q.931, Q.SIG, INAP, ISUP, TCAP, BSSAP, RANAP, MAP and CAP (mobile)
- Open APIs (JAIN, Parlay, and so on)

The IW-F exists when the MGC-F performs signaling interaction between different signaling networks (for example, SS7 and SIP). Examples of IW-F protocols include H.323/SIP and IP/ATM.

# Call Routing and Accounting Functions (R-F/A-F)

The R-F provides call routing information to the MGC-F, while the A-F collects call accounting information for billing purposes. The A-F can also have a broader role embodied by the common AAA functionality of authentication, authorization, and accounting in remote access networks. The primary role of both functions is to respond to requests from one or more MGC-Fs, directing the call or its accounting to terminating endpoints (other MGC-Fs) or services (AS-Fs).

R-F/A-F characteristics are as follows:

- Provides routing function for intra- and inter-network call routing (R-F).
- Produces details of each session for billing and planning purposes (A-F).

- Provides session management and mobility management.

- May learn routing information from external sources.

- May interact with the AS-F for the purposes of providing a service or feature to the user.

- May operate transparently to the other entities in the signaling path.

- Many R-Fs and A-Fs can be chained together in a sequential or hierarchical manner.

- The R-F/A-F is often integrated with the MGC-F. However, just as is the case with the AS-F, an integrated R-F/A-F/MGC-F can also request services of an external R-F/A-F.

- The A-F collects and emits per-call accounting information. The AS-F emits accounting information for enhanced services, such as conferences and premium information services.

- Applicable protocols for the R-F include ENUM and TRIP.

- Applicable protocols for the A-F include RADIUS and AuC (for mobile networks).

**SIP Proxy Server Function (SPS-F)** The most common embodiment of the R-F and A-F is as a SIP Proxy Server. For this reason, the ISC recognizes a separate *SIP Proxy Server Function* (SPS-F).

## Signaling Gateway Function (SG-F) and Access Gateway Signaling Function (AGS-F)

The SG-F provides a gateway for signaling between a VoIP network and the PSTN, whether SS7/TDM- or BICC/ATM-based. For wireless mobile networks, the SG-F also provides a gateway for signaling between an IP-based mobile core network and PLMN that is based on either SS7/TDM or BICC/ATM. The primary role of the SG-F is to encapsulate and transport PSTN (ISUP or INAP) or PLMN (MAP or CAP) signaling protocols over IP.

The AGS-F provides a gateway for signaling between a VoIP network and circuit-switched access network, whether V5- or ISDN-based. For wireless mobile networks, the AGS-F also provides a gateway for signaling between an IP-based mobile core network and PLMN that is based on either TDM or ATM. The primary role of the AGS-F is to encapsulate and transport V5 or ISDN (wireline), or BSSAP or RANAP (wireless) signaling protocols over IP.

SG-F characteristics are as follows:

- Encapsulates and transports PSTN signaling protocols (for example, SS7) using SIGTRAN to the MGC-F or another SG-F.

- For mobile networks, encapsulates and transports PSTN/PLMN signaling protocols (for example, SS7) using SIGTRAN to the MGC-F or another SG-F.

- The interface from the SG-F to the other entities is a protocol interface when the SG-F and MGC-F or other SG-F are not co-located (for example, SIGTRAN).

- One SG-F can serve multiple MGC-Fs.

- Applicable protocols include SIGTRAN, TUA, SUA and M3UA over SCTP.

AGS-F characteristics are as follows:

- Encapsulates and transports V5 or ISDN signaling protocols (for example, SS7) using SIGTRAN to the MGC-F.

- For mobile networks, encapsulates and transports BSSAP or RANAP signaling protocols (for example, SS7) using SIGTRAN to the MGC-F.

- The interface from the AGS-F to the other entities is a protocol interface when the AGF-F and MGC-F or other AGF-F are not co-located (for example, SIGTRAN).

- One MGC-F may serve many AGS-Fs.

- Applicable protocols include SIGTRAN, M3UA, IUA and V5UA over SCTP.

# Application Server Function (AS-F)

The AS-F is the application execution entity. Its primary role is to provide the service logic and execution for one or more applications and/or services.

AS-F characteristics are as follows:

- May request the MGC-F to terminate calls/sessions for certain applications (for example, voice mail or conference bridge).

- May request the MGC-F to reinitiate call features (for example, find me/follow me or prepaid calling card).

- May modify media descriptions using SDP.

- May control an MS-F for media handling functions.

- May be linked to web applications or have web interfaces.
- May have an API for service creation.
- May have policy, billing, and session log back-end interfaces.
- May interface with MGC-Fs or MS-Fs.
- May invoke another AS-F for additional services or to build complex, component-oriented applications.
- May use the services of an MGC-F to control external resources.
- Applicable protocols include SIP, MGCP, H.248, LDAP, HTTP, CPL and XML.
- Applicable open APIs include JAIN and Parlay.

Often the combination of the AS-F and the MGC-F provides enhanced call control services, such as network announcements, 3-way calling, call waiting, and so on. Rather than connecting the AS-F and MGC-F with a protocol, vendors often use an API between the AS-F and MGC-F when they are implemented in a single system. In this embodiment, the AS-F is known as a "Feature Server."

**Service Control Function (SC-F)**   The *Service Control Function* (SC-F) exists when the AS-F controls the service logic of a function. For this reason, the ISC recognizes a separate *Service Control Function* (SPS-F). Examples of SC-F protocols include INAP, CAP, and MAP; open APIs include JAIN and Parlay.

# Media Gateway Function (MG-F)

The MG-F interfaces the IP network with an access endpoint or network trunk, or a collection of endpoints and/or trunks. As such, the MG-F serves as the gateway between the packet and external networks, such as the PSTN, mobile network, and so on. For example, the MG-F could provide the gateway between an IP and circuit network (for example, IP to PSTN), or between two packet networks (for example, IP to 3G or ATM). Its primary role is to transform media from one transmission format to another, most often between circuits and packets, between ATM packets and IP packets, or between analog/ISDN circuits and packets as in a residential gateway.

MG-F characteristics are as follows:

- Always has a master/slave relationship with the MGC-F that is achieved through a control protocol such as MGCP or MEGACO.

- May perform media processing functions such as media transcoding, media packetization, echo cancellation, jitter buffer management, packet loss compensation, and so on.

- May perform media insertion functions such as call progress tone generation, DTMF generation, comfort noise generation, and so on.

- May perform signaling and media event detection functions such as DTMF detection, on/off-hook detection, voice activity detection, and so on.

- Manages its own media processing resources required to provide the functionality mentioned above.

- May have the ability to perform digit analysis based on a map downloaded from the MGC-F.

- Provides a mechanism for the MGC-F to audit the state and capabilities of the endpoints.

- Is not required to maintain the call state of calls passing through the MG-F; the MG-F only maintains the connection state of the calls it supports.

- A SIP phone is an MG-F and MGC-F in a single box.

- A SIP-capable gateway is an MG-F and MGC-F in a single box.

- "Hair pinning" of a call by the MG-F directed toward the source network may occur under control of the MGC-F.

- Applicable protocols include RTP/RTCP, TDM, H.248, and MGCP.

## Media Server Function (MS-F)

The MS-F provides media manipulation and treatment of a packetized media stream on behalf of any applications. Its primary role is to operate as a server that handles requests from the AS-F or MGC-F for performing media processing on packetized media streams.

MS-F characteristics are as follows:

- Support for multiple codecs and transcoding

- Support for control by multiple AS-Fs or MGC-Fs

- Support for multiple concurrent capabilities:
  - Digit detection
  - Streaming of tones and announcements (any multimedia file)
  - Algorithmic tone generation

- Recording of multimedia streams
- Speech recognition
- Speech generation from text
- Mixing (conference bridge)
- Fax processing
- Voice activity detection and loudness reporting
- Scripted combinations of the above

■ Performs under the control of an AS-F or MGC-F through a control protocol, with either tight coupling (resource control) or loose coupling (function invocation or scripts).

■ Applicable protocols include SIP, MGCP, and H.248.

# Media Gateway Controller Building Blocks

The *Media Gateway Controller* (MGC) is one of the key physical elements of a VoIP network. There have been many different implementations of the MGC and it is known by different names, including Softswitch, Call Agent, Call Controller, and others. Shown here are just some of the many possibilities available to equipment manufacturers and service providers under the ISC Reference Architecture (see Figure 14-3).

Most MGC systems today implement other functions in addition to the MGC-F. The other functions shown here (CA-F, IW-F, R-F, and A-F) could be collocated on the same physical platform or distributed across different systems that together constitute a total MGC solution. Of course, for load-balancing and availability purposes, the MGC can be implemented in a cluster of systems.

The functional building blocks in this MGC include the Connection Session Manager (MGC-F), Call Control & Signaling (CA-F), the Interworking/Border Connection Manager (IW-F) and the Access Session Manager (R-F/A-F). Other features implemented in the MGC shown include an Open Service Access Gateway, application Proxies, and OSS and OEM Agents. The Open Service Access Gateway and application Proxies together provide the discrimination and distribution of signaling and media for separate open and proprietary applications, respectively. The OSS and OEM Agents

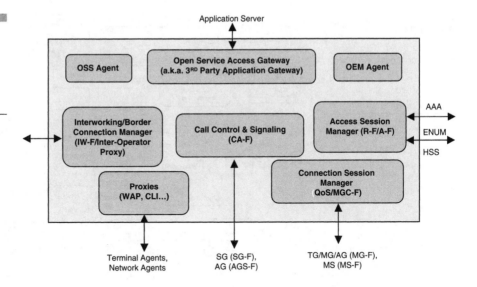

**Figure 14-3**
Media Gateway
Controller blocks in
the ISC Reference
Architecture
(Source: ISC)

interface to external OSS/OEM managers located in an operational support center for network management, service and network provisioning, maintenance, and so on.

## MGC Implementation Example

In this example implementation of an MGC, the MGC controls the Trunking Gateways, Media Gateways, and Media Servers using one or more of the media gateway control protocols through its MGC-F (see Figure 14-4). The MGC originates and terminates signaling messages to and from endpoints, other MGCs and external networks. Specifically, the Call Control & Signaling (CA-F) block performs this function. The Call Control & Signaling block also maintains the states of each call.

The example MGC system shown here also implements these other functions: Routing & Call Accounting (R-F/A-F), a SIP Proxy Server (SPS-F) and an Interworking/Border Connection Manager (IW-F). The MGC communicates via the Open Service Gateway Block (or application Proxy), with the application servers for services that are not native to the MGC. This could be done through various service control protocols and APIs, such as SIP, JAIN and Parlay.

**Figure 14-4**
Media Gateway
implementation in
the ISC Reference
Architecture
(Source: ISC)

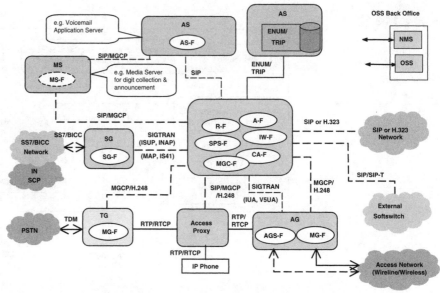

NOTE: The dashed lines represent the realization of information flows achieved by selecting specific protocol and interfaces to meet service requirements; this figure and its associated protocol choices is informative only.

# Network Examples

This section presents 10 examples of VoIP network configurations in a wide range of popular applications. Each example shows how the various functional entities might come together in physical devices and logical interactions to make up a VoIP network. In other words, each is just that—an *example* of one possible implementation, and no example is intended to represent a best or even a preferred approach.

## Wireline Network

The physical entities in this example consist of the MGC, *Applications Server* (AS), *Trunk Gateway* (TG), *Access Gateway* (AG), *Signaling Gateway* (SG) and a *Media Server* (MS) (see Figure 14-5).

The MGC in this example is carrying out the MGC-F, R-F, and A-F. The MGC terminates all the signaling (either directly or transported over IP) and carries out signaling interworking (for example, a SIP phone wanting

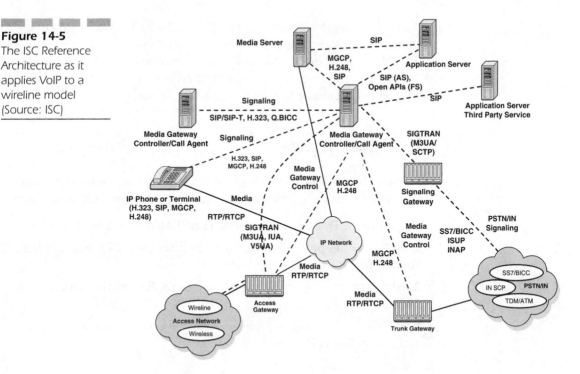

**Figure 14-5**
The ISC Reference
Architecture as it
applies VoIP to a
wireline model
(Source: ISC)

to signal to a PSTN network). It controls both the TG or AG for allocation of media resources. The MGC also authenticates and routes calls into the VoIP network and provides accounting information. Finally, the MGC interacts with other MGCs using either SIP/SIP-T, H.323, or Q.BICC.

The AS has the service logic for applications, such as voice mail. Calls requiring these functions can be either handed over by the MGC to the AS for service control, or the application server can provide the information required for the service logic execution to the MGC. The AS can control the MS directly or pass control of the MS to the MGC.

The TG terminates the physical media carrying the bearer (voice streams) from the PSTN, transcodes the bearer, and transports it over IP into the IP network. The TG is controlled by the MGC.

The AG serves as the interface between the IP network and any wireline or wireless access network. The AG transports signaling over IP to the MGC while transcoding the bearer and transporting it over IP to an IP endpoint or to a TG for sending it to a circuit or other packet network. The MG-F of the AG is controlled by the MGC.

The SG terminates the physical media for the PSTN signaling and transports the PSTN signaling to and from the MGC over IP.

The MS might perform tasks like announcements and digit collection, although the AG normally handles digit collection in most cases. The MS can be controlled either by the MGC, the AS, or both.

## All-IP Network

This example (see Figure 14-6) shows an all-IP network with the following components (which could be grouped physically in many different ways):

- ▪ **IP phones**  These that can support SIP, MGCP, and/or H.323.

- ▪ **H.323 Gatekeeper**  Domain administrator for H.323 based endpoints (R-F/A-F).

- ▪ **Media Server**  For providing advanced media functionality, such as IVR (MS-F).

- ▪ **SIP Proxy**  Routing and address translation entity for SIP-based endpoints (R-F/A-F).

**Figure 14-6**
The ISC Reference Architecture version of the converged network, that is, all IP (Source: ISC)

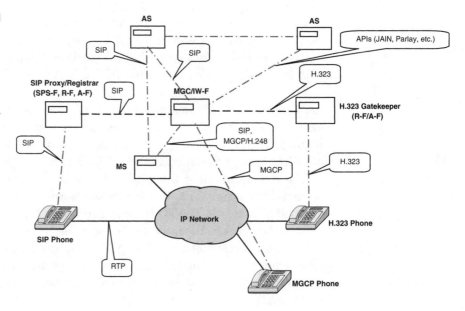

- **Application Server** Logical entity where services are hosted. One AS can also communicate with other Application Servers, which may implement services using JAIN, Parlay, and so on.

- **MGC/IW-F** In this architecture, this entity is used for interworking protocols; that is, for when SIP, H.323, and MGCP endpoints that need to communicate with each other. The MGC is also used for routing and accounting functions for controlling MGCP phones.

- **SIP Registrar** A SIP entity that keeps current locations for SIP devices.

The SIP phones update the SIP Registrar with their current address and use the proxy for routing. In cases where a SIP phone needs to communicate with non-SIP entities, it signals the MGC, which in this example handles the IW-F with the other protocols. Note that it is also possible to offer legacy phone features to SIP phones via the AS/MGC.

The MGCP phones contact the MGC for establishing and tearing down calls. The MGC also provides the R-F, A-F, and IW-F for the MGCP phone.

In the case of an H.323 phone, the H.323 Gatekeeper provides the R-F and A-F. The MGC provides protocol IW-F when required to communicate with non-H.323 networks.

In this case, media messages flow end-to-end between all media-capable devices using RTP/RTCP, regardless of the signaling protocol used. Hence, MGs are not required. An MG may be used if transcoding between different codecs is required, either on- or off-net.

The billing and accounting information may be exchanged between the gatekeeper, SIP proxy, and/or the MGC for overall accounting purposes, especially in case of multiprotocol calls.

The Application Servers may communicate directly with Media Servers or through the MGC to control the MS, as well as during the call to provide advanced media services like auto-attendant, IVR, conferencing, and so on.

## VoIP Tandem Switching

The example above shows a softswitch used for tandem trunk switching, which replaces the conventional PSTN Class 4 tandem switching (see Figure 14-7). The SG provides signaling transport conversion from SS7-based signaling protocols to IP SIGTRAN-based signaling protocols and transports. The TG, controlled by the MGC, provides trunking media (voice) connectivity in tandem from PSTN to IP to PSTN.

**Figure 14-7**
Reference
Architecture for
Class 4 (tandem)
replacement
(Source: ISC)

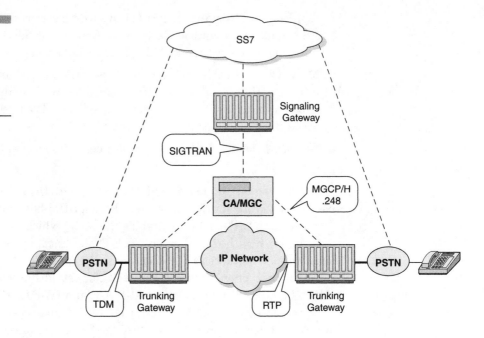

## POTS Carried over IP

This example shows the interconnection of a POTS phone to the PSTN via the IP network (see Figure 14-8). The POTS phone connects to the Residential Gateway (a type of Access Gateway). The RG performs the subscriber loop signaling and passes the signaling to the MGC using MGCP or the MEGACO protocol. The MGC in turn carries out signaling with the PSTN with the help of a Signaling Gateway. The RG digitizes and packetizes the analog voice and sends the digitized voice in RTP packets to the PSTN via the TG.

## Access Network (V5/ISDN) over IP

This example shows a V5 or GR303 and an ISDN-based access network (see Figure 14-9). The AG performs V5 or GR303 and ISDN signaling with the access network. The AG terminates the physical connection carrying the V5 or ISDN signaling and transports it over IP to the MGC using SIGTRAN

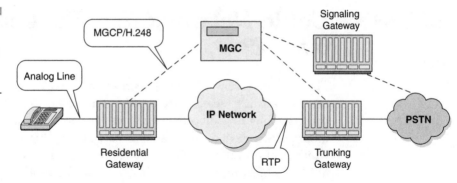

**Figure 14-8**
Reference
Architecture for POTS
to IP interface
(Source: ISC)

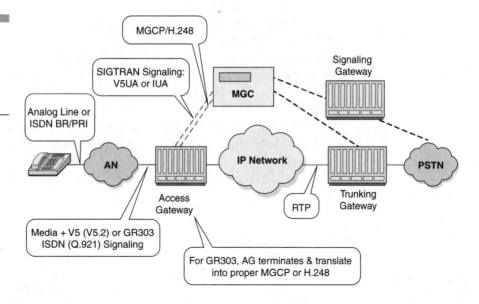

**Figure 14-9**
Reference
Architecture for
interfacing ISDN
and V5 (GR303)
(Source: ISC)

(V5UA or IUA). For GR303, the AG terminates the signaling and translates it into proper events of MGCP or MEGACO for transport to the MGC. The AG packetizes and possibly transcodes the voice stream from the access network and sends the voice packets to the TG using RTP packets. The TG converts the packetized voice to circuit voice and transmits it to the PSTN over the physical circuit-switched trunks.

# Cable Network (e.g. PacketCable™) over IP

This example shows the implementation of a VoIP network using a cable access network (see Figure 14-10). The cable modem at the customer premises has an embedded *Multimedia Terminal Adapter* (MTA) implemented in a device called an Access Gateway (or Residential Gateway), which connects to the POTS phones and any Ethernet-based devices. An MTA can be standalone or embedded into the cable modem; a standalone MTA interfaces to the cable modem via Ethernet. The MTA terminates the subscriber loop signaling to and from the POTS phone and communicates the signaling over IP (NCS or SIP) via the CMTS to the MGC. Network Control Signaling (NOS) is a modified form of MGCP. The MGC performs signaling with the PSTN using the Signaling Gateway. The MTA also terminates analog voice from the POTS phone, digitizes and packetizes the voice, and carries it over IP via the CM/CMTS cable network to the TG in RTP packets. The MGC controls the TG with TGCP (a modified form of MGCP).

To be fully PacketCable compliant, the MGC-F would also have to communicate with the CMTS using signaling such as COPS.

To ensure QoS when deploying VoIP over cable, the MGC communicates with the CMTS using the *Dynamic QoS* (DQoS) and COPS protocols.

**Figure 14-10**

Reference Architecture for converging voice and data via cable TV applications (Source: ISC)

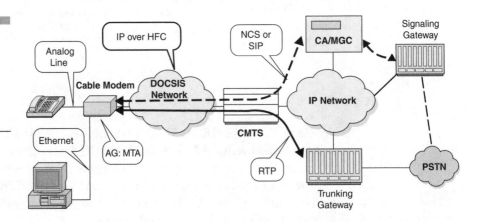

## VoDSL and IAD over IP

This example shows the implementation of a VoIP network using a DSL access network (see Figure 14-11). The *Integrated Access Device* (IAD) at the customer premises (also called Access Gateway or Residential Gateway, or asymmetric subscriber line termination unit) connects to the POTS phone and any Ethernet-based devices. The IAD provides the subscriber loop signaling and communicates the signaling over IP (MGCP or MEGACO) via the DSLAM to the MGC. The MGC carries out signaling with the PSTN using the SG. The IAD also digitizes and packetizes the voice and carries it over IP via the DSLAM to the TG in RTP packets.

## Wireless (3GPP R99 Special Case NGN)

This example shows how the wireless GSM/3G packet network connects to the PSTN or PLMN via the VoIP network (see Figure 14-12). The AG terminates the signaling (BSSAP in GSM or RANAP in 3G) from the *radio access network* (RAN) over E1/T1/ATM interfaces. It transports the signaling messages over IP to the MSC Server using SIGTRAN. The MSC and GMSC servers carry out the same functionality as an MGC, including signaling with the PSTN via the SG, or to the PLMN via the GMSC server. The

**Figure 14-11**
VoIP and DSL in
Reference
Architecture
(Source: ISC)

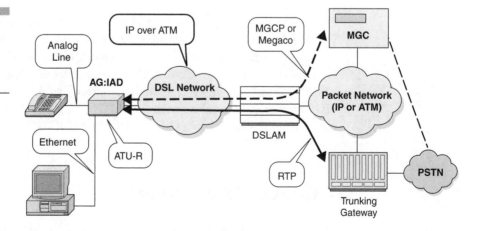

**Figure 14-12**
Softswitch in wireless
networks under
Reference
Architecture
(Source: ISC)

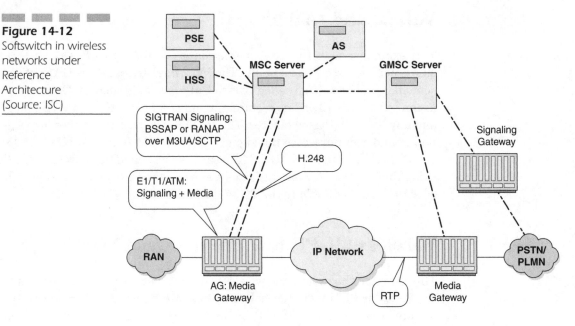

media from the RAN is terminated on the AG, transcoded and transported to the TG as RTP packets.

## Wireless (3GPP R2000 General Case all IP)

This example shows how the wireless 3G packet network connects to the PSTN or PLMN via the VoIP network (see Figure 14-13). The SGSN/GGSN passes the signaling (BSSAP in GPRS or RANAP in 3G) from the RAN over IP to the *Multimedia Server Call Server/Application Server* (MMCS/MMAS), which provides the same functionality as an MSC Server. The MGC carries out signaling with the PSTN or legacy PLMN via the SG. The media from the RAN is passed through the SGSN/GGSN to the MG as RTP packets.

## WCDMA Mobile Network

This example shows a total WCDMA network architecture in the circuit-switched domain, with reference to 3GPP, where an IP-based core network

**Figure 14-13**
Softswitch in 3G
wireless networks
under Reference
Architecture
(Source: ISC)

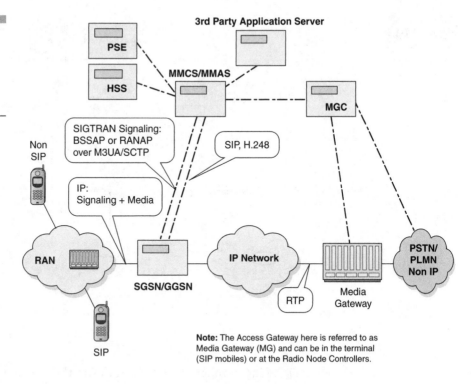

is used (see Figure 14-14). This situation is similar to the Wireless R99 example above, but includes more complete protocols. In addition, it shows how an MS, controlled by the MSC Server, can provide simple announcements and under the control of an Application Server that delivers value-added services, such as voice messaging, push-to-talk, and conferencing.

# Conclusion

This chapter explored the history, the present, and the future of softswitch. Far too many softswitch success stories exist to be contained in one book. The chief prognosticator for softswitch over the next two years will be how ubiquitous softswitch becomes as a long-distance solution and, most importantly, how widespread softswitch, in whatever form, takes traffic away from Class 5 switches.

**Figure 14-14**
Reference
Architecture for 3GPP
networks on an IP
network (Source: ISC)

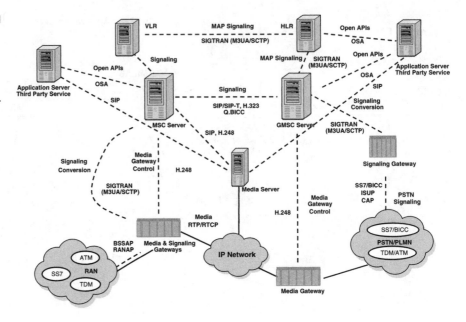

The growing number of success stories in the industry coupled with reference architecture, protocols, and standards will assuage fears of service providers contemplating a move to softswitch technology. Eventually, a critical mass will be reached and softswitch will replace both Class 4 and 5 switches.

# CHAPTER 15

# Conclusion and Prognostications

Although this book has focused on comparing softswitch solutions to legacy Class 4, Class 5, and *private branch exchange* (PBX) architectures, the reality is that softswitch solutions are something entirely new and should not be compared to legacy systems on a port-for-port or feature-by-feature basis. The brave new world of telecommunications is based on the *Internet Protocol* (IP) and offers voice as one of a multitude of services. The telecommunications infrastructure in developed economies will evolve from legacy to converging to converged networks. Capitalism dictates that entirely new infrastructures will rise to compete with legacy architectures for a share of the multibillion dollar international telecommunications market. Given that softswitch solutions are cheaper, smaller, simpler, and more convenient to use, they could well be the platforms of choice in gaining those market shares.

Despite the fact that softswitch technologies are entirely new, it is necessary to describe softswitch in comparison to legacy *Time Division Multiplexing* (TDM) architectures. In researching and working with softswitches, I've repeatedly found there are no problems, only solutions that a talented engineering team can build into their network. Unlike legacy telephone networks, softswitch offers a modularity of interoperable components that enable service providers to "mix and match" to build a network that meets their goals. Table 15-1 compares Class 4 switches and some softswitch products that might replace them.

**Table 15-1**

Softswitch meets or exceeds the performance parameters of Class switches

| Vendor and product | DS0s/ rack | BHCAS | Reliability | SS7 | MOS | Price per DS0 |
|---|---|---|---|---|---|---|
| Nortel DMS-250 (Class 4) | 2,688 | 800,000 | 99.999 | Y | 4.0 | $100 |
| Lucent 4ESS (Class 4) | 2,688 | 700,000 | 99.999 | Y | 4.0 | $100 |
| Convergent ICS2000 (Softswitch) | 108,864 | 1,500,000 | 99.9994 | Y | 4.0 | $25 |
| SONUS GSX9000 (Softswitch) | 24,000 | 2,000,000 | 99.999 | Y | 4.0 | $25 |
| Nuera Nu-Tandem (Softswitch) | 6120 | 480,000 | 99.999 | Y | 4.83 | $75 |

# Softswitch and the PSTN Metrics of Performance

Table 15-1 compares legacy Class 4 switches against Class 4 replacement softswitches. The Class 4 replacement softswitches appear to meet or exceed the performance parameters of the legacy Class 4 switches. These are the criteria that most service providers have for selecting new platforms.

Softswitch architecture is more scalable than Class 4 at both ends of the spectrum. Some softswitch solutions can put 100,000 DS0s in one seven-foot rack where a comparable Class 4 solution (Nortel's DMS-250, for example) requires over a dozen seven-foot racks to deliver the same density. More importantly, a softswitch solution is superior to a Class 4 switch, for example, in its ability to sale down rather than up. A softswitch solution can scale down to one DS0 in the case of an IP phone. The smallest scale of a DMS-250, for example, is 480 DS0s. This gives a service providers greater flexibility in deployment to the markets served.

Softswitch solutions are potentially more reliable than Class 4 or Class 5, which boast five 9s of reliability. This merely describes the switch only and not the *Public Switched Telephone Network* (PSTN) as a whole. *Voice over IP* (VoIP) is essentially a voice over a data network. Many are the data networks that are *high availability* (HA) with the five 9s. This describes the network as a whole and is not limited to one network element, the switch in this case. The PSTN as a network will never achieve five 9s as each central office, for example, constitutes a *single point of failure* (SPOF). As many subscribers found to their dismay on September 11, 2001, there is no back up to any given central office.

Softswitch solutions potentially meet or exceed Class 4 and 5 in terms of *quality of service* (QoS). The primary concern for service providers and subscribers is the need to match the voice quality of the PSTN, which rates a *Mean Opinion Score* (MOS) of 4.0. Some softswitch solutions exceed 4.0. To do this, service providers must engineer the latency of voice traffic to a level that meets or beats the PSTN. Multiple mechanisms can be engineered into an IP network to deliver adequate QoS, including *Differentiated Services* (DiffServ), *Resource Reservation Setup Protocol* (RSVP), and *Multiprotocol Label Switching* (MPLS). Components of a softswitch solution, such as media gateways, can also be engineered to minimize latency.

Another objection to VoIP and softswitch solutions is the perception that softswitch solutions cannot deliver the 3,500 features of a Class 5 switch in North American markets. No service provider in that market needs to offer 3,500 features to any subscriber or market of subscribers. In the Class 5 market, a service provider can obtain new features only from the switch vendor. Historically, a service provider would have to spend millions of dollars and wait years for features to be delivered by their switch vendor. Softswitch solutions use open standards that allow the rapid creation and rollout of new features. These features can come from a softswitch vendor, a third-party software vendor, or the service provider can write the feature themselves.

Service providers have also been concerned about interfacing IP networks and the PSTN, especially in regards to the flow of signaling between those two networks. The use of *Signaling System 7* (SS7) is necessary for the delivery of features to subscribers. SS7 can be routed over IP networks using *Signaling Transport* (SIGTRAN) and other emerging mechanisms. It is possible that SS7 will be replaced by simpler, more efficient means of signaling such as *Session Initiation Protocol* (SIP).

# Softswitch and the PSTN Infrastructure

Given the lower cost of acquisition and *Operations, Administration, Maintenance, and Provisioning* (OAM&P) provided by softswitches, new market entrants will want to leverage the lower barrier to entry into telephony markets. These combined factors could potentially give new market entrants advantages over incumbents. Table 15-1 does not address the issue of features for the Class 4 switch, because features are the function of a Class 5 switch. However, given its capability to interface with Class 5 feature sets via SS7, a Class 4 replacement softswitch can deliver all the same features of a Class 4 switch. A Class 4 replacement softswitch, in the definition of disruptive technology, is cheaper, simpler, smaller, and more convenient to use than the Class 4 switch.

Each of the three TDM switch platforms, Class 4, Class 5, and PBX, face different threats of disruption or deconstruction from different softswitch solutions. These softswitch platforms threaten the legacy TDM switch platforms in that they are simpler and more convenient to use.

Perhaps the greatest threat to the TDM Class 5 switch is the IP PBX. Some 70 percent of corporate telephony is interoffice traffic. If that traffic were to move entirely to the corporate *wide area network* (WAN) or any other non-PSTN network, demand for Class 5 switched services and tangentially, Class 5 switches, would plummet (a trend already under way).

The growth in demand for SIP-based telephony products (IP PBXs and PC-based soft phones) will enable enterprises to move their voice traffic away from the PSTN. The inclusion of SIP in Microsoft's Windows XP may accelerate this trend.

The use of access switching where TDM traffic moves to an IP network at the neighborhood pedestal also replaces the Class 5 as the central office is bypassed. This could prove to be a cost-effective alternative to building or maintaining a central office. It would also prove to be an attractive strategy for new market entrants in offering local telephone service, especially in greenfield (that is, new construction neighborhoods) applications.

The use of IP phones in conjunction with softswitch in a distributed architecture could prove to be disruptive as it requires only that the subscriber has an IP phone and that multiple service providers are involved in the completion of a call (IP service provider, telephony service provider, and feature service provider). This could prove disruptive to incumbent TDM service providers with legacy networks that are limited in the services, applications, and features they can offer. An accelerant to this scenario would be ubiquitous broadband to the home and small business. Wireless technologies such as *wireless fidelity* (Wi-Fi) will make rapid deployment possible for networks that compete with the PSTN.

## Alternatives to the Telephone Company

Another misleading assumption in this discussion is that only telephone companies can deliver voice services and that they have no incentive to invest in new technologies when their legacy networks are delivering revenue. This has been true for over 100 years. Telephone companies, given the economies of scale necessary to deliver telephone service in most markets, have historically enjoyed a monopoly in their markets. Privatization and the introduction of competition in these markets has emphasized the need to find platforms that allow competitors to offer comparable service for less investments in infrastructure.

Historically, a telephone network consisted of three elements: transport, switching, and access. In the United States, the transport market was opened by the *Memorandum of Final Judgement* (MFJ) in 1984. The recent bandwidth glut has opened this market further for competitive service providers. The switching and access portions of the network were supposed to be opened by the Telecommunications Act of 1996. To date, less than 10 percent of U.S. households have any choice of local service providers. Softswitch solutions offer competitive service providers a means of bypassing the switching facilities of the *incumbent local exchange carrier* (ILEC). Access facilities of the ILEC can be bypassed with wireless, cable TV, or fiber optic cable (known as *fiber to the home* [FTTH]).

## Alternative Private Service Providers

Alternatives to the ILECs' switching and access facilities open the telephone service market to a host of alternative service providers. Primarily, *competitive local exchange carriers* (CLECs) can offer a competitive alternative to ILEC services using softswitch and other technologies. Other alternatives are made possible by softswitch technologies. Wireless ISPs can deliver voice service via their data services using softswitched VoIP. Cable TV companies can bundle voice services in their video and data services. New technologies are coming on the market that enable the delivery of IP-based services via power lines.

## Alternative Public Service Providers

An intriguing possibility is the deployment of *metro area networks* (MANs) by municipalities. Ashland, Oregon; Spokane, Washington; and Longmont, Colorado are examples of cities that have deployed and own fiber optic, IP, municipally owned MANs. These municipalities operate the MAN, which is open to any and all service providers to deliver services to subscribers on the MAN, including Internet, voice, and video.

In the early 1990s, one American cable TV executive promised subscribers 500 channels of video services via his cable network. A decade later no such line up of services has materialized. An IP MAN makes it possible to offer subscribers 500 different service providers who in turn can offer an almost unlimited lineup of content in video, Internet, and voice. An IP connection could be thought of as a utility not unlike a road, power line, water line, or sewer pipe connected to a residence or business. Although many would argue against "government competing with business," few businesses can compete with municipalities in delivering roads, water, and sewer services for homes and businesses. Perhaps the chief reason less than 10 percent of U.S. households have access to broadband is that the cost of building the infrastructure for broadband negates an advantageous return on investment. If the cost of building broadband networks were shouldered by municipalities with an eye on the economic benefits to the community, there would be a more rapid rollout of such services.

Municipalities that built such networks could recoup the expense of building those networks by charging service providers (ISPs, telephone companies, and video service providers) for access to those networks, not unlike states charging license fees and gas taxes for roads. In some cases, it is possible that a municipality could replace property tax revenue with the revenue it receives from telecommunications services, thus eliminating property taxes. This would be especially true if the municipality were the ISP, phone company, and cable TV company receiving hundreds of dollars per month per household in gross telecommunications revenue.

What does all this have to do with softswitch? Of the three services, data, voice, and cable TV, voice offers the greatest revenues. Softswitch, because it is cheaper and more convenient to use, simplifies the delivery of voice services. Revenues from voice services or the perception of value in low-cost voice services by the subscriber (residential or business) might be the deciding factor in deploying such a network.

# The End of the PSTN as We Know It?

Figure 15-1 forecasts a plausible scenario for the end of the PSTN as we know it. Telephony is already only one of many applications available via an IP connection. The use of VoIP can save businesses considerable sums of money in telephony costs. The rollout of broadband and wireless technologies can only accelerate this trend.

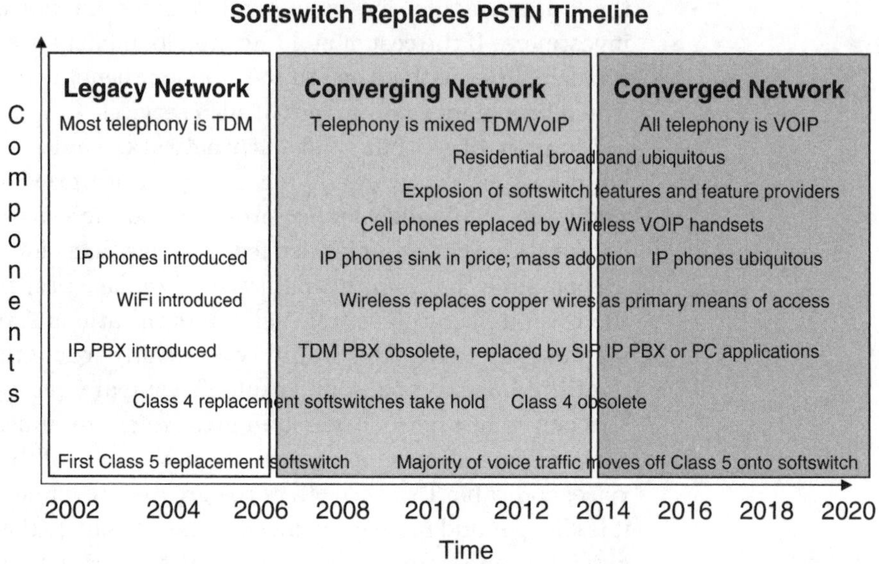

**Figure 15-1**
The timeline for the
convergence of voice
and data networks,
resulting in the
replacement of the
PSTN

**Softswitch Replaces PSTN Timeline**

| | **Legacy Network** | **Converging Network** | **Converged Network** |
|---|---|---|---|
| | Most telephony is TDM | Telephony is mixed TDM/VoIP | All telephony is VOIP |
| | | Residential broadband ubiquitous | |
| | | Explosion of softswitch features and feature providers | |
| | | Cell phones replaced by Wireless VOIP handsets | |
| | IP phones introduced | IP phones sink in price; mass adoption  IP phones ubiquitous | |
| | WiFi introduced | Wireless replaces copper wires as primary means of access | |
| | IP PBX introduced | TDM PBX obsolete, replaced by SIP IP PBX or PC applications | |
| | Class 4 replacement softswitches take hold    Class 4 obsolete | | |
| | First Class 5 replacement softswitch    Majority of voice traffic moves off Class 5 onto softswitch | | |

C o m p o n e n t s

2002    2004    2006    2008    2010    2012    2014    2016    2018    2020

Time

# ACRONYMS LIST

The Glossary is based on the "Reference Architecture" document by International Softswitch Consortium 2002.

| | |
|---|---|
| **3G** | Third generation |
| **3GPP** | 3G Partnership Project (UMTS) |
| **AAA** | Authentication, authorization, and accounting (IETF) |
| **A-F** | (Call) accounting function (ISC) |
| **ACELP** | Algebraic Code-Excited Linear Prediction |
| **ADPCM** | Adaptive Differential Pulse-Code Modulation |
| **ADSL** | Asymmetric Digital Subscriber Line |
| **AF** | Assured forwarding |
| **AG** | Access gateway |
| **AGS-F** | Access Gateway Signaling Function (ISC) |
| **AIN** | Advanced Intelligent Network |
| **AN** | Access Network |
| **ANSI** | American National Standards Institute |
| **API** | Application programming interface |
| **AS** | Application server |
| **AS-F** | Application server function (ISC) |
| **ASN.1** | Abstract Syntax Notation 1 |
| **ATM** | Asynchronous Transfer Mode |
| **ATU-R** | ADSL Terminal Unit-Remote |
| **AuC** | Authentication center (GSM) |
| **BG** | Border gateway |
| **BGP** | Border Gateway Protocol |
| **BICC** | Bearer Independent Call Control (ITU Q.1901) |
| **BSSAP** | Base Station Subsystem Application Part (GSM, 3GPP) |
| **CA** | Call agent |
| **CA-F** | Call Agent Function (ISC) |
| **CAMEL** | Customized Applications for Mobile Network Enhanced Logic (GSM) |
| **CAP** | CAMEL Application Part (GSM, 3GPP) |

| | |
|---|---|
| **CAS** | Channel Associated Signaling |
| **CATV** | Cable television |
| **CBR** | Constant bit rate |
| **CBR** | Content-based routing |
| **CCS** | Common Channel Signaling |
| **CDMA** | Code Division Multiple Access |
| **CDR** | Call Detail Record |
| **CELP** | Code-Excited Linear Prediction |
| **CIC** | Circuit Identification Code |
| **CLEC** | Competitive Local Exchange Carrier |
| **CLI** | Common Language Infrastructure |
| **CM** | Cable modem |
| **CMTS** | Cable Modem Termination System (DOCSIS, PacketCable) |
| **COPS** | Common Open Policy Protocol (IEFT RFC 2748) |
| **CPE** | Customer premise equipment |
| **CPL** | Call-Processing Language |
| **DiffServ** | Differentiated Services |
| **DNS** | Domain name system |
| **DOCSIS** | Data Over Cable System Interface Specification |
| **DPC** | Destination point code |
| **DQoS** | Dynamic quality of service |
| **DSL** | Digital subscriber line |
| **DSLAM** | DSL access multiplexer |
| **DSP** | Digital signal processor |
| **DTMF** | Dual tone/multiple frequency |
| **EC** | Echo cancellation |
| **EF** | Expedited forwarding |
| **ENUM** | E.164 Numbering (IETF RFC 2916) |
| **ER** | Explicit route |
| **ESP** | Enhanced service provider |
| **ETSI** | European Telecommunications Standards Institute |
| **FCS** | Frame check sequence |
| **FDM** | Frequency Division Multiplexing |
| **FEC** | Forward Equivalence Class |

| | |
|---|---|
| **FS** | Feature server |
| **FTP** | File Transfer Protocol |
| **GGSN** | Gateway GPRS System Node (GPRS, 3GPP) |
| **GK** | Gatekeeper |
| **GMSC** | Gateway Mobile Services Switching Center (GSM, 3GPP) |
| **GPRS** | General Packet Radio Service |
| **GSM** | Global System for Mobility |
| **HFC** | Hybrid fiber/cable |
| **HLR** | Home Location Register (GSM, 3GPP) |
| **HSS** | Home Subscriber System (3GPP) |
| **HTTP** | Hypertext Transport Protocol (IETF) |
| **IAD** | Integrated access device |
| **IETF** | Internet Engineering Task Force |
| **ILEC** | Incumbent Local Exchange Carrier |
| **IN** | Intelligent Network |
| **INAP** | Intelligent Network Application Protocol |
| **IP** | Internet Protocol |
| **ISDN** | Integrated Services Digital Network |
| **ISUP** | Integrated Services Digital Network User Part (SS7) |
| **ITSP** | Internet Telephony Service Provider |
| **ITU** | International Telecommunications Union |
| **ITU-T** | International Telecommunication Union—Telecommunications Standardization Sector |
| **IUA** | ISDN user adaptation |
| **IVR** | Interactive Voice Response |
| **IW-F** | Interworking function (ISC) |
| **JAIN** | Java Application Interface Network |
| **LAN** | Local area network |
| **LDAP** | Lightweight Directory Access Protocol (IEFT) |
| **LEC** | Local exchange carrier |
| **LMDS** | Local Multipoint Distribution Service |
| **LSP** | Label-switched paths |
| **LSR** | Label switching router |
| **LSSU** | Link status signal unit |

| | |
|---|---|
| **M2UA** | MTP2-User Adaptation Layer |
| **M3UA** | MTP3 User Adaptation (IETF SIGTRAN) |
| **MAP** | Mobile application part (GSM, 3GPP) |
| **MC** | Multipoint controller |
| **MCU** | Multipoint control unit |
| **MEGACO** | Media gateway control (IETF RFC 3015 or ITU H.248) |
| **MG** | Media gateway |
| **MGC** | Media gateway controller |
| **MGC-F** | Media gateway controller function (ISC) |
| **MGCP** | Media Gateway Control Protocol (IETF RFC 2705) |
| **MG-F** | Media gateway function (ISC) |
| **MMAS** | Mobile multimedia application server (3GPP) |
| **MMCS** | Mobile multimedia call server (3GPP) |
| **MOS** | Mean Opinion Score |
| **MPLS** | Multiprotocol label switching |
| **MS** | Media server |
| **MSC** | Mobile Services Switching Center (GSM, 3GPP) |
| **MS-F** | Media server function (ISC) |
| **MSU** | Message signal unit |
| **MTA** | Multimedia terminal adapter (PacketCable) |
| **MTP** | Message transfer part |
| **MTU** | Maximum transmission unit |
| **NCS** | Network call/control signaling (PacketCable) |
| **NEBS** | Network Equipment Building System |
| **NIST** | National Institute of Standards and Technology |
| **NGN** | Next-generation network |
| **OEM** | Original equipment manufacturer |
| **OSA** | Open service access (3GPP) |
| **OSI** | Open System Interconnection |
| **OSPF** | Open shortest path first |
| **OSS** | Operational Support System |
| **PAM** | Pulse amplitude modulation |
| **PBX** | Private branch exchange |
| **PC** | Personal computer |

| | |
|---|---|
| **PC** | Point code |
| **PCM** | Pulse code modulation |
| **PDD** | Post-dial delay |
| **PHB** | Per hop behavior |
| **PLMN** | Public Land Mobile Network (3GPP, UMTS) |
| **POTS** | Plain old telephone service |
| **PRI** | Primary rate interface |
| **PSE** | Personal service environment (3GPP) |
| **PSTN** | Public Switched Telephone Network |
| **QoS** | Quality of service |
| **RAN** | Radio Access Network |
| **RANAP** | Radio access network application part (3GPP) |
| **RAS** | Registration, admission, and status |
| **R-F** | (Call) routing function (ISC) |
| **RFC** | Request for Comment (IETF) |
| **RG** | Residential gateway |
| **Rseq** | Response sequence |
| **RSVP** | Resource Reservation Protocol (IETF) |
| **RTCP** | Real-Time Control Protocol (IETF) |
| **RTP** | Real-Time Transport Protocol (IETF RFC 1889) |
| **RTSP** | Real-Time Streaming Protocol |
| **SAP** | Session Announcement Protocol |
| **SC-F** | Service control function (ISC) |
| **SCCP** | Signaling connection control part |
| **SCP** | Service control point |
| **SCTP** | Stream Control Transmission Protocol |
| **SDP** | Session Description Protocol |
| **SID** | Silence insertion descriptor |
| **SG** | Signaling gateway |
| **SG-F** | Signaling gateway function (ISC) |
| **SGSN** | Serving GPRS system node (GPRS, 3GPP) |
| **SIGTRAN** | Signaling Transport (IETF M3UA, IUA, SUA, V5UA Drafts) |
| **SIP** | Session Initiation Protocol (IETF) |
| **SIP-T** | SIP for Telephony (IETF Draft) |

| | |
|---|---|
| **SLA** | Service level agreement |
| **SNMP** | Simple Network Management Protocol |
| **SPS-F** | SIP Proxy Server Function (ISC) |
| **SS7** | Signaling System 7 |
| **SSP** | Service signaling point |
| **STP** | Signal transfer point |
| **SUA** | SCCP user adaptation (IETF SIGTRAN) |
| **TA** | Terminal adapter |
| **TCAP** | Transaction capability application part (SS7) |
| **TCP** | Transmission Control Protocol |
| **TDM** | Time Division Multiplexing |
| **TG** | Trunk or trunking gateway |
| **TGCP** | Trunking Gateway Control Protocol |
| **TLA** | Three-letter acronym |
| **TOS** | Type of service |
| **TRIP** | Telephony Routing over IP (IETF RFC 2871) |
| **TUA** | TCAP user adaptation (IETF SIGTRAN) |
| **UMTS** | Universal Mobile Telecommunications System |
| **URL** | Uniform resource locator |
| **V5UA** | V5 user adaptation (IETF SIGTRAN) |
| **VAD** | Voice Activity Detection |
| **VBR** | Variable bit rate |
| **VLR** | Visitor location register (GSM, 3GPP) |
| **VoATM** | Voice over ATM |
| **VoDSL** | Voice over DSL |
| **VoIP** | Voice over IP |
| **VXML** | Voice Extensible Markup Language |
| **WAN** | Wide area network |
| **WAP** | Wireless Application Protocol |
| **WCDMA** | Wideband Code Division Multiple Access |
| **XML** | Extensible Markup Language |

# INDEX

## Symbols

4ESS tandem switch, 12
5ESS switches, 12
802.11b standard, 289

## A

A-law, 20
access, last mile, 289
access alternatives, 289
  fiberless optics, 290
  Wi-Fi, 289
access switching, 124-125
access, PSTN
  DTMF function, 10
  handsets, 10
ACK message, SIP, 91, 94
Active/Standby HA systems, 139
addressing, signaling, 21
AF service (assured forwarding), DiffServ,
    166-167
AGS-F, ISC reference architecture, 314
AIN (Advanced Intelligent Network), 27
  intelligent peripherals, 29
  SCPs, 28-29
  SS7, 28
alerting, signaling, 21
all-IP VoIP network example, 322-323
amplifiers, FDM, 16
APIs (application program interfaces), 52
  softswitch, 202-203
    CORBA, 205
    JAIN, 204
    Parlay, 203
    XML, 206
    XML.NET, 207

application servers, 52, 209
  interactions, 216-217
  media server interface, 216
  SIP interface, 212, 214-215
architecture
  Class 4/5 switches, 13
  MPLS, 170
  SIP clients, 89
  softswitch. See softswitch architecture.
architecture of ISC, 208
  application servers, 209
  media servers, 209
ARMIS (Automatic Reporting and
    Management Information
    System), 134
AS-F, ISC reference architecture, 315
ASN.1 (Abstract Syntax Notation), 99
ASPs (application service providers), 120
associations, SCTP, 184
ATM, 34-35
  long-distance voice transport, 63
  VCIs (virtual circuit identifier), 35
audio codecs, H.323, 74
availability, calculating, 137
availability of power, 148

## B

bandwidth, QoS, 156
bandwidth saving, softswitch, 231
barriers to entry, VoIP gateways, 227
basic services
  DMS-250, 30
  PSTN features, 197
best-effort service, RSVP, 83
best-effort traffic, RSVP, 83
BGP (Border Gateway Protocol), 81

interautonomous system routing, 81

intra-autonomous system routing, 81

passthrough autonomous system
routing, 82

routing, 82

BHCAs (Busy Hour Call Attempt), 13, 118

BICC (Bearer Independent Call Control), 78

billing preservation, HA systems, 141

BN (annual benefit), 243

broadband, 288

  cable TV, 288

  consumer benefits, 294

  consumer surplus, 291

  economic benefits, 294

  lower cost services, 293

  multiple voice connections, 292

  NDRBN (National Defense Residential
Broadband Network), 295

  producer surplus, 291

  telephone services, 267-268

  usage of existing telephone network, 292

  Wi-Fi, 289

BYE requests, SIP, 95

bypass operators, 226


**C**

CA-F, ISC reference architecture, 313

cable TV

  access alternatives, 289

    fiberless optics, 290

    Wi-Fi, 289

  broadband, 288

cable VoIP network example, 326

calculating availablity, 137

calculating unavailability, 137

CALEA (Communications Assistance for
Law Enforcement Act), 196, 218-220

call agents, 76-77

call control, 50

Call Control and Signaling Plane, ISC
reference architecture, 311

call control entities, SIP interface, 212,
214-215

call preservation, HA systems, 141

call setup, SIP

  redirect server, 95

  via proxy server, 92-94

  via registrar, 95

  via UASs, 89

call signaling, H.225, 75

call stateful proxies, SIP, 92

call-processing power

  softswitch, 118

    scalability, 115

CAROT (Centralized Automatic Reporting
on Trunks), 13

carrier grade net present value models, 257

carrier-grade gateways, 48-49

CAS (Channel-Associated Signaling), 22

cascaded MCs, H.323 Version 2, 104

centralized physical architecture model, 211

Centrex, 15

channel banks, TDM, 16

chunks, SCTP, 185

Cisco IOS software reliability case study,
147-148

CLASS (Custom Local Area Signaling
Service), 29

  features, 196

  services, 108

Class 4 discounted DS0 present value
model, 253

Class 4 leasing versus softswitch cost
structure, 250

Class 4 market disruption, 277, 281, 283

Class 4 purchase versus softswitch, 252

Class 4 replacement applications, 120

Class 4 replacement switches, 56-57, 60

Class 4/5 switches

  QoS, 33

reliability, 31
scalability, 33
PSTN, 12
architecture, 13
components, 13-14
Class 5 replacement applications, 121
Class 5 replacement switches, 59-60
CLECs (Competitive Local Exchange Carriers), 2
clients, SIP, 89-90
cluster mode HA systems, 139
$C_N$ (operating costs), 242
codecs, 18-19
G.711, 20
G.723.1 ACELP, 21
G.728 LD-CELP, 20
G.729, 21
comfort noise generators, 160
companding, 17
competitive local loop environment, obstacles, 2
component reliability, 143
compression, VoIP, 69
conferencing, H.323, 104
configurations, SIP, 97
connections, 77
consumer benefits from broadband, 294
consumer surplus, broadband, 291
control signaling, H.245, 75
controlled load service, QoS, 164
controlled-delay service, RSVP, 84
converged architecture, 60
converged broadband networks, economic benefit, 291
converged networks, 266, 288
converging architecture, 60
CORBA, softswitch API, 205
cost effectiveness, softswitch, 229
bandwidth saving, 231
lower barrier to entry, 232
OAM&P, 230

purchase cost, 230
cost per busy hour call attempt, 261
cost savings
broadband, 293
integrated access, 293
CPE (customer premise equipment), PSTN, 10

**D**

data flows, RSVP, 83
decompression, VoIP, 70
deconstruction, 272
service providers, 273-275
vendors, 276
delay
fixed, 164
IP routers, 157
QoS, 156
queuing, 164
VoIP gateways, 158
delay-sensitive traffic, RSVP, 83
densities, media gateways, 115
developing economies, 265
DiffServ (Differentiated Service), 165
AF service, 166-167
DSCP markers, 165
EF service, 165
GoS (grade of service), 165
PHBs (per-hop behaviors), 165
SLAs, 165
disruption of legacy telecom value network, 277
Class 4 market, 277, 281-283
PBX market, 284
disruptive technologies, 7
distributed architecture, 40
access, 42
IP handsets, 43-44

media gateways, 47, 49

PC to PC/PC to phone, 42

VoIP gateways, 46

distributed architecture advantages, softswitch, 237

distributed physical architecture model, 212

DLECs (data local exchange carriers), 120

DMS-250, 13-14

basic services, 30

enhanced services, 30

DS0 support, softswitch scalability, 115

DS1, 16

DSCP markers, DiffServ, 165

DSPs (digital signal processors), media gateway density, 118

DTMF function (Dual-Tone Multifrequency), PSTN access, 10


**E**

E1, 16

E911, softswitch, 218

echo, QoS, 156

economic advantages, softswitch, 7

economic benefit of converged broadband networks, 291

economic benefits, converged networks, 266

economic benefits of broadband, 294

economic issues for softswitch, 239

economics, enterprise softswitch applications, 262

economics of softswitch, 226-227

EF service (expedited forwarding), DiffServ, 165

encoding, 18

end-to-end delay, QoS, 157

endpoint identifiers, 77

endpoints, 76

enhanced services

DMS-250, 30

PSTN features, 197

enterprise gateways, 48

enterprise softswitch application economics, 262

ENUM (Telephone Number Mapping), 96

error codes, SIP, 105

error detection/correction, VoIP, 70

ESS systems, 12

extensibility, 105


**F**

factors in QoS, voice quality, 156

Fast Start, H.323, 101

FCC (Federal Communications Commission), 4

FDM (frequency division multiplexing), 16

features, softswitch, 6

FECs (Forwarding Equivalence Classes), MPLS, 168

feedback, H.323/SIP comparisons, 104

fiber cable, 36

fiber optic transmission, 36

fiberless optics, 290

FIT (failures in time), 135

five 9s of availability, 31

five 9s of reliability, softswitch, 135, 138

fixed delay, 164

footprint, softswitch, 116, 233-236

functional entities, ISC reference architecture, 311

functional planes, ISC reference architecture, 309


**G**

G.711 codecs, 20

G.723.1 ACELP codecs, 21

G.728 LD-CELP codecs, 20

G.729 codecs, 21
gatekeepers, 51
   H.323, 74
   zones, 51
gateway control protocols, VoIP, 76
gateways
   H.323, 73
   minimizing QoS impairments, 159
geographical redundancy, softswitch, 143
GoS (grade of service), DiffServ, 165
GR-2914-CORE, 151
GR-454-CORE, 151
guaranteed bit-rate service, RSVP, 83
guaranteed service, QoS, 164
GUIs, IP handsets, 45

## H

H.225, 72
   call signaling, 75
   RAS (registration, admission, and
      status), 75
H.245, 72
   control signaling, 75
H.323, 71-72
   audio codecs, 74
   conferencing, 104
   Fast Start, 101
   gatekeepers, 74
   gateways, 73
   limited extensibility, 106
   MCUs, 74
   multimedia network interworking, 73
   poor scalability, 100-102
   RTCP, 75
   RTP, 75
   server processing, 103
   SIP comparison, 98
   SIP simplicity, 99-100
   terminals, 73

   video codecs, 74
   VoIP signaling, 72
   zones, 74
H.323 Version 2, cascaded MCs, 104
HA systems (Highly Available), 137
   Active/Standby, 139
   call/billing preservation, 141
   cluster mode systems, 139
   hot switchover, 141
   redundancy, 139
handsets, PSTN access, 10
headers, RTP, 85
high-availability power/environment
      recommendations, softswitch, 149
history of softswitch, 304
hot standby, 141
hot switchover
   HA systems, 141
   rapid execution, 141
   rapid fault detection, 141
human error, reliability, 150-151
hybrid codecs, 18

## I

i-Mode, 223
IEEE 802.11b standard, 289
IEEE 802.1p specification, 162
IGRP (Interior Gateway Routing
      Protocol), 80
ILECs (Incumbents Local Exchange
      Carriers), 2
implementing MGCs, 319
improving QoS on network, 161
IN (Intelligent Network), 23
in-channel signaling, 22
IN/AIN (Intelligent Network/Advanced
      Intelligent Network), 198-199
INAP (Intelligent Network Application
      Part), 79, 108

individual interface downtime,
     softswitch, 146
integrated access, cost savings, 293
intelligent peripherals, AIN, 29
interautonomous system routing,
     BGP, 81
international VoIP traffic, 228
interworking H.323 and SS7, 193
intra-autonomous system routing, BGP, 81
IntServ (Integrated Services), 162
INVITE command, SIP, 90, 94
IP addresses, IP phones, 44
IP backbone providers, 63
IP Centrex
     with Class 5 switch architecture, 54
     with softswitch architecture, 55
IP handsets, 43-44
     GUIs, 45
     IP addresses, 44
IP networks, 61-62
     MPLS, 167-168
IP PBXs, 53
IP routers, delay source, 157
IP telephony, lower cost telephone
     services, 293
IPv6, 85
ISC (International Softswitch
     Consortium), 208
     application servers, 209
     media servers, 209
     reference architecture, 307-308
        AGS-F, 314
        AS-F, 315
        CA-F, 313
        Call Control and Signaling Plane, 311
        functional entities, 311
        functional planes, 309
        Management Plane, 311
        MG-F, 316
        MGC-F, 312-313

        MS-F, 317-318
        R-F/A-F, 313-314
        Service and Application Plane, 311
        SG-F, 314
        Transport Plane, 310
ISUP (ISDN User Part), 79
     encapsulation in SIP, 191
     SS7, 27, 179
IUA (ISDN Q.921-User Adaptation
     Layer), 189
IXCs (interexchange carriers, 34


## J-K

JAIN, softswitch API, 204
jitter, QoS, 156
jitter buffer, VoIP, 70
key systems, PSTN, 15
killer apps, 222


## L

label switching, MPLS, 169
lasers, fiber optic, 36
last mile, 289
LD-CELP codecs (Code-Excited Linear
     Predictor), 20
LDAP (Lightweight Directory Access
     Protocol), 89
LDPs (label-distribution protocols),
     MPLS, 171
LECs (local exchange carriers), 34
legacy architecture, 60
limited extensibility, H.323, 106
links, SS7, 23-25
LNP (local number portability), 79
local loop outages, 3-4
location servers, SIP, 96

long-distance bypass operators, VoIP gateway switches, 226

long-distance voice transport, ATM, 63

lower barrier to entry, softswitch cost effectiveness, 232

LSAs (link-state advertisements), OSPF, 80

LSPs (label-switched paths), MPLS, 168, 171

LSRs (label-switching routers), MPLS, 170

# M

M2UA (MTP2 User Adaptation Layer), 186

M3UA (MTP3 User Adaptation Layer), 186-187

Management Plane, ISC reference architecture, 311

MAP (Mobile Application Part), 79

MAP messages (Mobile Application Part), SS7, 27

market drivers, softswitch, 51

MCUs (Multipoint Control Units), H.323, 74

media gateways, 46-49, 181
  densities, 115
  POPs, 114
  scalability, 117
  super nodes, 115

media servers, 209, 216

MEGACO, 71

MF (R1 Multifrequency), 22

MFJ (Memorandum of Final Judgement), open transport competition, 288

MG-F, ISC reference architecture, 316

MGC-F, ISC reference architecture, 312-313

MGCP (Media Gateway Control Protocol), 71, 76

MGCs (media gateway controllers), 50, 76, 181, 318

implementation, 319

softswitch call-processing, 118

microwave systems, 35

midsize long-distance service providers, net present value models, 249

MIME content, SIP messages, 108

mobility feature, SIP, 94

models for midsize long-distance service providers net present value, 249

models for softswitch net present value, 241-247

modularity, SIP, 107

Moore's Law
  models for softswitch net present value, 242-243
  rate of change decrease, 140
  reduced cost of redundancy, 139

MOS (Mean Opinion Score), 173

MPLS (Multiprotocol Label Switching), 167
  architecture, 170
  FECs (Forwarding Equivalence Classes), 168
  IP networks, 167-168
  label switching, 169
  LDPs (label-distribution protocols), 171
  LSPs (label-switched paths), 168, 171
  LSRs, 170
  LSRs (label-switching routers), 170
  traffic engineering, 171-172

MS-F, ISC reference architecture, 317-318

MTBF (mean time between failure), 135

MTP (Message Transfer Part), SS7, 26, 178

MTTR (mean time to repair), 135

Mu-law, 20

multiple voice connections, broadband, 292

multiplexers, TDM, 16

multipoint connections, 77

multiport interface downtime, softswitch, 146

mux, TDM, 16

## N

Napster, 222

NDRBN (National Defense Residential Broadband Network), 295

NEBS (Network Equipment Building Standards), 32, 144
  softswitch reliability, 4-5

net present value calculations, softswitch solutions, 6, 239-240

net present value models
  carrier grade, 257
  midsize long-distance service providers, 249
  softswitch, 241-247
    generating greater revenue, 255
    versus discounted Class 4, 253
  summary, 259, 261

network connectivity, softswitch, 143

NGDLC (Next-Generation Digital Loop Carrier), 124

NHLFE (Next Hop-Level Forwarding Entry), MPLS LSRs, 170

noise suppression, VoIP, 69

non-interoffice free long distance, 262

nonuniform quantization, 17

Nortel DMS-250, 13

numerical error codes, SIP, 105


## O

OADMs (optical add/drop multiplexers), 36

OAM&P (Operations, Administration, Maintenance, and Provisioning)
  4ESS, 13
  softswitch cost effectiveness, 230

obstacles, competitive local loop environment, 2

OC-N (optical carrier level N), 36

OCC (opportunity cost of capital), 242

OCX (optical carrier level), 36

one to five 9s, 135

OpenTel case study, VoIP gateways, 227-228

optical transmission systems, 35

origins of VoIP, 68

OSPF (Open Shortest Path First), LSAs (link-state advertisements), 80

out-of-band signaling, 22

outside plant, PSTN, 10


## P

packages, 77

packet loss, QoS, 156

PacketCable, 326

packetization, VoIP, 69

packets
  RTP, 84-85
  SCTP, 185

PAM (pulse amplitude modulation), 17

parameters, voice codecs, 19

parlay, softswitch API, 203

passthrough autonomous system routing, BGP, 82

PATH message, RSVP, 163

payloads, RTP packets, 85

PBXs (private branch exchanges), 14-15
  market disruption, 284
  TDM (time division multiplexing), 16
  voice mail systems, 15

PC-to-PC service, 42

PC-to-phone service, 42

PCM (pulse code modulation), 17
  companding, 17
  encoding, 18
  PAM, 17
  quantization, 17

PDD (postdial delay), SIP, 91

PEP (Protocol Extensions Protocol), 106

performance metrics, Class 4/5 switches
  QoS, 33
  reliability, 31
  scalability, 33

PHBs (per-hop behaviors), DiffServ, 165
phone-to-phone VoIP, 43
physical architecture models
  centralized, 211
  distributed, 212
physical endpoints, 77
PINT (PSTN to Internet Interworking),
  96, 192
point codes, SS7 signaling points, 23-24
point-to-point connections, 77
poor scalability, H.323, 100-102
POPs (points of presence)
  media gateways, 114
  VoIP gateways, 227
POTS over IP VoIP network example, 324
power availability, 148
power draw advantages, softswitch, 236
power outages, PSTN, 149
predictive service, RSVP, 84
producer surplus, broadband, 291
protocol intermediation, 51
protocol stack, SS7, 26
protocols
  SIGTRAN, 183
  SS7, 79
  VoIP, 70
    routing, 79
    signaling, 71
proxy routing, SIP, 108
proxy servers, SIP, 89, 92-94
PSQM (Perceptual Speech Quality
  Measurement), 160, 173
PSTN (Public Switched Telephone
  Network), 2
  access, 10
  access alternatives, 289
    fiberless optics, 290
    Wi-Fi, 289
  basic services, 197
  Centrex, 15
  Class 4/5 architecture, 13
  Class 4/5 components, 13-14

Class 4/5 switches, 12
CPE (customer premise equipment), 10
disruption by softswitch, 334
enhanced services, 197
FDM (frequency division
  multiplexing), 16
interoperation with VoIP, 178
lack of redundancy, 4
outside plant, 10
PBXs, 14
power outages, 149
redundancy, 139
signaling, 21
SPOFs, 141
switching, 11-12
TDM (time division multiplexing), 16
transport, 34
  ATM, 34-35
vulnerability, 132, 134
PTT organizations (Post, Telegraph, and
  Telephone), 226
public service provider alternatives,
  336-337
purchase cost, softswitch, 230

**Q**

QoS (Quality of Service), 5, 156
  Class 4/5 switch performance
    metrics, 33
  controlled load service, 164
  end-to-end delay, 157
  guaranteed service, 164
  improving on network, 161
  minimizing gateway impairments, 159
  MOS (Mean Opinion Score), 173
  PSQM, 173
  RSVP, 83
  voice quality, 156
quantization, 17
queuing delay, 164

# R

R-F/A-F, ISC reference architecture, 313-314
RADIUS (Remote Access Dial-In User Service), 89
rapid execution, hot switchover, 141
rapid fault detection, hot switchover, 141
RAS (registration, admission, and status), H.225, 75
rate-sensitive traffic, RSVP, 83
redirect servers, SIP, 89, 95
redundancy
  HA systems, 139
  PSTN, 139
redundant softswitch hardware nodes, 140
reference architecture, ISC, 307-308
  AGS-F, 314
  AS-F, 315
  CA-F, 313
  Call Control and Signaling Plane, 311
  functional entities, 311
  functional planes, 309
  Management Plane, 311
  MG-F, 316
  MGC-F, 312-313
  MS-F, 317-318
  R-F/A-F, 313-314
  Service and Application Signaling Plane, 311
  SG-F, 314
  Transport Plane, 310
regenerators, TDM, 16
registrars, SIP, 89, 95
regulatory issues, softswitch, 6-7, 238-239
reliability, 4-5, 135
  Class 4/5 switch performance metrics, 31
  components, 143
  human error, 150-151
  power availability, 148
  PSTN vulnerability, 132-134
  softswitch specs, 144-146
  software, 148
  Telcordia specs, 135-136
repeaters. *See* regenerators.
requests, SIP, 90
require headers, SIP, 105
residential gateways, 47
residential media gateways, 122-123
response codes, SIP, 93
RESV message, RSVP, 163
RFC (R2 Multi-Frequency Compelled), 22
RINGING message, SIP, 94
RINGING response, SIP, 91
RIP (Routing Information Protocol), 80
robustness, SCTP, 186
routing, BGP, 82
routing protocols, VoIP, 71, 79
RSVP (Resource Reservation Protocol), 83, 89, 162-163
  data flows, 83
  PATH message, 163
  QoS, 83
  RESV message, 163
  traffic types, 83
RTCP (RTP Control Protocol), 75, 85
RTP (Real-Time Protocol), 84
  H.323, 75
  headers, 85
  packets, 84-85
  transport protocol, 88
RTP (Real-Time Transport Protocol), 71

# S

SAP (Session Announcement Protocol), 88
scalability, 5
  Class 4/5 switch performance metrics, 33
  H.323, 102
  media gateways, 117

scalability of softswitch, 114
  call-processing power, 115
  DS0 support, 115
scaling down, 129
  for Class 5 bypass applications, 121
  for Class 5/CO bypass from residence,
     122-123
  for Class 5/CO bypass from the
     business, 126
  technical issues, 127-128
  to replace Class 4 switches, 120
  via residential media gateways, 122-123
  via softswitch, 119
SCCP (Signaling Connection Control Part),
    SS7, 27
SCE, 200-201
SCE (service creation environment),
    softswitch, 198
SCPs (service control points)
  AIN, 28-29
  SS7, 24
SCTP (Stream Control Transmission
    Protocol), 184
  associations, 184
  chunks, 185
  packets, 185
  robustness, 186
  streams, 185
SDH (Synchronous Digital Hierarchy),
    36-37
SDP (Session Description Protocol), 88
server processing, H.323, 103
servers, SIP, 89-90
Service and Application Plane, ISC
    reference architecture, 311
service APIs, 216
service providers
  alternatives, 336
  deconstruction, 273-275
services, SIP, 108
SG-F, ISC reference architecture, 314

signaling, 5-6, 50
  CAS (Channel-Associated Signaling), 22
  CCS (Common Channel Signaling), 22
  PSTN, 21
  SS7 links, 23
signaling gateways, 52, 182
signaling in VoIP networks, 181
signaling network architecture, 190
signaling points, SS7, point codes, 23-24
signaling protocols
  SIP, 88
  VoIP, 71-72
SIGTRAN (Signaling Translation), 96,
    180-183
silence suppression, 160
SIP (Session Initiation Protocol), 71, 88
  ACK message, 91, 94
  architecture, clients, 89
  BYE requests, 95
  call control-application server interface,
    212, 214-215
  call setup
    via proxy server, 92, 94
    via redirect server, 95
    via registrar, 95
    via UASs, 89
  CLASS services, 108
  clients, 89-90
  configurations, 97
  error codes, 105
  extensibility, 105
  H.323 comparison, 98
  INVITE command, 90, 94
  ISUP encapsulation, 191
  location servers, 96
  MIME content, 108
  mobility feature, 94
  modularity, 107
  PDD (postdial delay), 91
  proxy routing, 108
  requests, 90

Require headers, 105
response codes, 93
RINGING message, 94
RINGING response, 91
servers, 89-90
services, 108
signaling protocol, 88
simplicity compared to H.323, 99-100
standards, 96-97
transactions, 90
TRYING message, 94
TRYING response, 90
UACs (UA clients), 90
UAs, 89
UA servers, 89
Windows XP, 109-110
SIP URLs, 88
SIP-T (SIP for Telephones), 192
SLAs, DiffServ, 165
small footprint, softswitch, 233-234, 236
social rewards of softswitch, 296
  housing, 299
  no commuting, 296-300
softswitch, 4
  access switching, 124-125
  APIs, 202-203
    CORBA, 205
    JAIN, 204
    Parlay, 203
    XML, 206
    XML.NET, 207
  broadband telephone service, 267-268
  CALEA, 218-220
  call control, 50
  call-processing power, 118
  Class 4 purchase case study, 252
  Class 4 replacement switches, 56-57, 60
  Class 5 replacement switches, 59-60
  compared with PSTN performance
    metrics, 333
  cost effectiveness, 229

  bandwidth saving, 231
  lower barrier to entry, 232
  OAM&P, 230
  purchase cost, 230
  cost per busy hour call attempt, 261
  cost structure versus Class 4 leasing, 250
  distributed architecture
    advantages, 237
  E911, 218
  economic advantages, 7
  enterprise application economics, 262
  features, 6
  five 9s of reliability, 138
  footprint, 116
  geographical redundancy, 143
  high-availability power/environment
    recommendations, 149
  improving quality of life, 300-301
  IP PBXs, 53
  IP Centrex, 54-55
  market drivers, 51
  net present value models, 239-247
    carrier grade, 257
    discounted Class 4, 253
    generating greater revenue, 255
  network connectivity, 143
  power draw advantages, 236
  public service provider alternatives,
    336-337
  QoS, 5, 156
  redundant hardware nodes, 140
  regulatory issues, 6-7, 238-239
  reliability, 4-5
  reliability specs, 144, 146
  residential media gateways, 122-123
  scalability, 5
  scaling down, 119
    for Class 5 bypass applications, 121
    for Class 5/CO bypass from the
      business, 126
    to replace Class 4 switches, 120

SCE (service creation environment), 198-201
service provider alternatives, 336
signaling, 5-6, 50
small footprint, 233-234, 236
social rewards, 296
  family life, 300
  housing, 299
  no commuting, 296-298
telephone company alternatives, 335
voice-activated web interface, 222
web provisioning, 221
softswitch and SS7, 176-177
softswitch architecture, 40
  access, 42
  IP handsets, 43-44
  media gateways, 47, 49
  PC to PC/PC to phone, 42
  VoIP gateways, 46
softswitch case studies
  Class 5 replacement, 306
  long-distance Class 4 replacement, 305-306
softswitch economics, 226-227
softswitch history, 304
softswitch scalability, 114
  call-processing power, 115
  DS0 support, 115
software reliability, 148
software reliability case study, Cisco IOS, 147-148
SOHO gateways, 47
SONET (Synchronous Optical Network), 36-37
source codecs, 18
sources of delay
  IP routers, 157
  VoIP gateways, 158
SPF routing algorithm, 80
SPIRITS (Services in the PSTN/IN Requesting Internet Services), 96, 192

SPOFs (single points of failure), 32, 134
  PSTN, 141
  PSTN vulnerability, 134
SS7 (Signaling System 7), 23
  AIN, 28
  ISUP, 179
  ISUP (ISDN User Part), 27
  links, 23-25
  MAP messages (Mobile Application Part), 27
  MTP, 178
  MTP (Message Transfer Part), 26
  protocol stack, 26
  protocols, 79
  SCCP, 179
  SCCP (Signaling Connection Control Part), 27
  SCPs (service control points), 24
  signaling gateways, 52
  signaling points, point codes, 23-24
  SSPs (service switching points), 23
  STPs (signal transfer points), 24
  TCAP, 179
  TCAP (Transaction Capabilities Applications Part), 27
  TUP (Telephone User Part), 27
SS7 and softswitch, 176-177
SS7-VoIP interworking, 180
SSCP, SS7, 179
SSPs (service switching points), SS7, 23
standards, SIP, 96-97
stateful proxies, SIP, 92
stateless proxies, SIP, 94
Stowger switches, 12
STPs (signal transfer points), SS7, 24
streams, SCTP, 185
STS signals (synchronous transport signal), 36
STS-1 (synchronous transport signal level 1), 37
SUA, 187-188

summary of net present value models, 259-261

sunk costs, VoIP gateways, 227

super nodes, 115

supervision, signaling, 21

switches, ATM, 34-35

switching, PSTN, 11-12

   Class 4/5 architecture, 13

   Class 4/5 components, 13-14

   Class 4/5 switches, 12

synchronous optical transmission, 36

## T

T-1, 16

T-Carrier, 16

tandem switching VoIP network example, 323

TCAP (Transaction Capabilities Application Part), SS7, 27, 79, 179

TCP (Transmission Control Protocol), 71

TCP/IP protocol stack, 63

TDM

   multiplexers, 16

   networks, 64

   regenerators, 16

technical issues for scaling down, 127-128

Telcordia

   procedural error documents, 151

   reliability specs, 135-136

telecom bust, 4

Telecommunications Act of 1996, 2, 65

   opening of switching/access, 289

telephone company alternatives, 335

terminals, H.323, 73

total service downtime, softswitch, 145

traffic engineering, MPLS, 171-172

traffic types, RSVP, 83

transactions, SIP, 90

transmitting telephone calls, 16

transport, 34-35, 60

Transport Plane, ISC reference architecture, 310

transport protocols

   RTP, 88

   VoIP, 84

TRIP (Telephony Routing over IP), 97

TRYING message, SIP, 94

TRYING response, SIP, 90

TSI memory matrix (time slot interchange), 12

TUA (TCAP-User Adaptation Layer), 189

TUP (Telephone User Part), SS7, 27

## U

UACs (UA clients), SIP, 90

UAs (user agents), SIP, 89

UASs (user agent servers), SIP, 89

UDP (User Datagram Protocol), 71

ULL (unbundled local loop), 6

unavailability, calculating, 137

UPSs (Uninterruptible Power Supplies), 149

## V

V5/ISDN IP VoIP network example, 324

VAD (voice activation detection), 160

VCIs (virtual circuit identifiers), ATM, 35

vendors, deconstruction, 276

video codecs, H.323, 74

virtual endpoints, 77

vocoders, 18

VoDSL and IAD over IP VoIP network example, 327

voice codecs, 19

voice digitization, PCM, 17

voice mail systems, 15

voice quality
  PSQM, 160
  QoS, 156
  VoIP, 173
voice transport, 60
voice-activated web interface, 222
VoIP (Voice over Internet Protocol), 4, 44
  compression, 69
  decompression, 70
  error detection/correction, 70
  gateway control protocols, 76
  jitter buffer, 70
  noise suppression, 69
  origins, 68
  packetization, 69
  phone-to-phone, 43
  protocols, 70
  QoS, 156
  routing protocols, 71, 79
  signaling, 181
  signaling protocols, 71-72
  transport protocols, 84
  voice quality, 173
VoIP gateways, 46
  barriers to entry, 227
  delay source, 158
  international long-distance operators, 229
  replacing Class 4 switches, 120
  sunk costs, 227
VoIP interoperation with PSTN, 178
VoIP networks
  all-IP, 322-323
  cable, 326
  POTS over IP, 324
  tandem switching, 323
  V5/ISDN, 324
  VoDSL and IAD over IP, 327
  wireless, 327-328
  wireline, 320-321
VoIP-SS7 interworking, 180
vulnerability of PSTN, 132, 134
VXML (Voice XML), 6, 108

**W**

waveform codecs, 18
WCDMA mobile networks, 328
web provisioning, 221
Wi-Fi (wireless fidelity), 289
Windows Messenger, Windows XP, 110-111
Windows XP
  SIP, 109-110
  Windows Messenger, 110-111
wireless VoIP network example, 327-328
wireline VoIP network example, 320-321

**X-Z**

XML, softswitch API, 206
XML.NET, softswitch API, 207
zones
  gatekeepers, 51
  H.323, 74

# ABOUT THE AUTHOR

*Photo by Lucinda Welch*

Frank Ohrtman
President, Softswitch Consulting
http://www.softswitchconsulting.com
720-839-4063

Frank Ohrtman has many years experience in sales of VoIP and softswitch platforms. His career in VoIP began with selling VoIP gateway switches for Netrix Corporation to long distance bypass carriers. He went on to promote softswitch solutions for Lucent Technologies (Qwest Account Manager) and Vsys (Western Region Sales Manager). The genesis of this book lies in answering customer objections to VoIP and, tangentially, softswitch technologies.

Mr. Ohrtman learned to perform in-depth research and write succinct analyses during his years as a Navy Intelligence Officer (1981-1991). He is a veteran of U.S. Navy actions in Lebanon (awarded Navy Expeditionary Medal), Grenada, Libya (awarded Joint Service Commendation Medal) and the Gulf War (awarded National Defense Service Medal). Mr. Ohrtman holds a Master of Science degree in Telecommunications from Colorado University College of Engineering (master's thesis: "Softswitch As Class 4 Replacement—A Disruptive Technology") and a Master of Arts degree in International Relations from Boston University.